houseplants for all

Photo: Ron Goh

houseplants for all

HOW TO FILL ANY HOME
with HAPPY PLANTS

DANAE HORST

Houghton Mifflin Harcourt
Boston New York 2020

For my mom, whose love of plants started me on a journey I could never have imagined would lead me here.

For information about permission to reproduce selections from this book, write to trade.permissions@hmhco.com or to Permissions, Houghton Mifflin Harcourt Publishing Company, 3 Park Avenue, 19th Floor, New York, New York 10016.

hmhbooks.com

Library of Congress Cataloging-in-Publication Data is available

ISBN: 978-0-358-37994-2 (hbk)
ISBN: 978-0-358-37996-6 (ebk)

Edited and designed by Girl Friday Productions
www.girlfridayproductions.com
Book design by Katy Brown

Printed in China

SCP 10 9 8 7 6 5 4 3 2 1

contents

INTRODUCTION

PLANTS HAVE DEEP ROOTS IN MY LIFE. When I was a child, my mother and I would make frequent outings to the two nurseries in the small Wyoming town where I grew up. As we wandered through those misty greenhouses and rows of plants, I'd ask the names of the ones that caught my eye while hunting for fallen blooms to carefully stow in my pocket and take home. From those greenhouse visits of my youth to the now frequent ones I make sourcing plants for my shop, Folia Collective, the wonder of plants has not worn off for me.

Unfortunately, enjoying plants and keeping them happy do not always have a direct correlation, as I discovered when I started to collect my own houseplants in earnest while living in Seattle. Making mistake after mistake (most of which I didn't even know were mistakes until years later) and watching my plants struggle and/or die, I began to wonder if my mother's "green thumb" had skipped a generation with me. When my husband and I moved to Los Angeles, I packed up the handful of plants I had managed to keep alive, nurturing the tiniest seed of hope that maybe they'd be happier in a sunnier locale. While an environment with more plentiful sun and warmer temperatures did improve my odds, I still had a lot to learn—only two of the plants I brought to L.A. survived. Whether from a place of folly or just sheer determination, I continued to try caring for new plants and allowed my love for them to motivate me to learn about them.

When I began to choose plants based on what they needed rather than just how I wanted my home to look, I found my plants seemed happier and healthier. This was the breakthrough I needed. As I took the time to learn more about their care, my mistakes became evident, and I grew more confident in my plant-parenting abilities.

Shortly after, a new job working for the design blog/studio Jungalow plunged me into photo styling with not just a few houseplants, but *all* the plants. The more I worked with plants, the more I wanted to learn about them, and the more I learned, the more I wanted to work with them. At the same time, I began to notice so many people constantly lamenting that they were "plant killers," and I knew that the myth of the "black thumb" would eventually discourage them, as it had me. Justina Blakeney, Jungalow's founder, put the perfect words to what my experience had taught me: "There's no such thing as a green thumb; just get to know your plants and love them."

Driven by my desire to help dispel the "black thumb" myth, I began to study houseplant care even more and started to help friends find the right plants for both their space and their lifestyle. One day while plant shopping for a friend's new home, she said, "You should do this as a job." At the time I thought, "Yeah, right—I wish this was a job," but that comment stuck with me like a little sprout and grew until I couldn't ignore it any longer. The idea of the plant styling studio and boutique plant shop that would become Folia Collective began to take root.

Now, nearly four years after starting Folia, every person I've talked plants with and every plant project I've worked on have reinforced what I believe to be true:

PLANTS MAKE ANY SPACE FEEL ALIVE. Whether it's a staged home about to go on the market, a small business looking to add vitality to its workspace, or a budding plant lover's apartment welcoming new foliage, every time we take a space from "before plants" to "after plants," their power is evident.

THERE'S NO SPECIAL MAGIC NEEDED TO KEEP PLANTS HAPPY AND HEALTHY. It's truly about knowing what plants need and choosing the ones that are best suited to your space, your lifestyle, and your personality.

STYLING YOUR HOME WITH HAPPY & HEALTHY PLANTS MEANS STYLING *WITH THE PLANTS IN MIND.* I understand the draw to certain popular plants, especially when social media teases us with images of perfect plants in immaculate homes. If you only choose a plant based on its popularity or style factor, you might get a few cute photos out of it, but if you can't provide what it needs, the plant will not be happy in the long run.

PLANTS GIVE US A MUCH-NEEDED CONNECTION TO THE NATURAL WORLD. Even in the middle of a concrete jungle, a little potted plant can draw a line from our home to the actual jungle it came from. In a time when more and more people are moving to urban environments, plants usher in a small piece of the wonder of the natural world, remind us of our place in it, and hopefully inspire us to take better care of it.

In the pages of this book, you'll find houseplant inspiration from real homes in varied climates and locales; but more importantly, you'll develop a better understanding of the kind of environment you can offer your plants so they can thrive and bring you joy for years to come. First, you'll discover how to assess your own home for essentials like light and humidity. Next, you'll meet the right plants for each of the common home environments and learn about each plant's needs. Finally, you'll pick up fundamental skills for plant care and how to solve common plant problems.

I believe plants are for all of us. No matter your budget, your tastes, or your lifestyle, you can infuse your home with the vitality that plants provide.

danae

the right plants for you

Most of us have been there: You see a plant at the store, fall in love with it, and bring it home, only to watch it struggle and perhaps even die a slow death, all the while wondering, "What am I doing wrong? Why can't I make you happy?" Eventually you succumb to the notion that you aren't a "plant person" or buy into that old "black thumb" myth. The truth is, anyone can be a plant person! The first step is to learn the essentials about what plants need to be happy and healthy, and then choose the plants whose needs match what you, and your environment, can provide for them.

We'll do a deep dive into everything plants need a little later in the book, but first, some key ingredients to plant care every successful plant person needs to understand: light and humidity.

LIGHT

Light is to plants as food is to animals. *Photosynthesis*, the process through which plants convert light into sugar, creates the energy that plants need to survive. Water and nutrients are also important, but without light, nearly all plants eventually die. Light is the most essential need a plant has, and yet it's one of the most misunderstood and under-considered aspects of plant care.

MISCONCEPTION: *I don't get direct sun, so I can't have plants.*
TRUTH: Most indoor plants don't require direct sun, and many won't tolerate it for more than a few hours. In the afternoon, direct sun (in the Northern Hemisphere) is especially strong and can scorch plants' leaves. Nearly all plants do best in bright indirect light (see pages 4–6).

MISCONCEPTION: *I have a light bulb near a plant. That's enough light, right?*
TRUTH: Most household light bulbs don't have enough of the spectrum of light that plants need, so apart from a fluorescent light bulb (which low-light plants can usually work with), or a type of light called a grow light (see page 56), a light bulb alone isn't enough light for a plant.

MISCONCEPTION: *I have a room with no windows. A low-light plant will be fine there.*
TRUTH: While some low-light plants can tolerate zero natural light for a time, they will eventually die of starvation in a room with no natural light. A grow light setup is the only way to keep plants happy long-term in a room without a window.

MISCONCEPTION: *I have a large window, so plants on the other side of the room are getting lots of light.*
TRUTH: Once you understand the way light travels through a room, you'll notice that even in rooms with large windows, areas far away from the windows or in a corner aren't receiving much light at all. Plants that require bright light need a spot right near the window, preferably in front of it where the light is strongest.

Direct Sun vs. Indirect Light

In the world of plant care, "direct sun" occurs when the actual sun (not just light) is visible through the window and so the sun's rays are hitting your plants. Indirect light occurs when ambient light is present but the sun is not directly visible through the windows that let the light in. Window treatments that allow most of the light through can also filter the direct sun into indirect light. Around our shop, I often explain the difference by asking this question: "Imagine the plant has eyes. Can it see the sun?" If your answer is *yes*, the plant gets direct sun. If your answer is *no*, but the room is still very bright, the plant gets bright indirect light.

Direct sun comes in at an angle. This angle changes depending on the time of year, the building position, and the architectural style. The area closest to the window receives the brightest light but also gets hit with the sun's rays directly, which can burn delicate leaves. As you move farther from the window, the strength of the light drops off.

Indirect light appears to illuminate the room in more of an arc shape as it comes through the window. Light still tapers off as you move farther from the window, and you'll notice the corner areas closest to the window are just as dark as the areas farthest from the window.

Understanding the difference between these two types of light is important since many plants will be overwhelmed by direct sun. You'll need to consider the kind of light an area receives when selecting your plants for that area. We'll dig into this more in Section Two.

You'll also encounter certain light terms when looking at care information for plants. Understanding these terms better will help you choose plants that are suited for your space and care for them properly.

DIRECT SUN OR FULL SUN: The plant needs to be in the path of the sun (remember, if the plant had eyes, could it see the actual sun?) and get 4 or more hours of direct sun each day.

BRIGHT INDIRECT LIGHT: The plant needs a bright spot, usually within a few feet of a window, where there are 6+ hours of light, but not direct sun, each day.

MEDIUM OR LOW LIGHT: There are very few plants that require this kind of light indoors, but if you encounter this on a tag, that plant can tolerate a spot farther away from a window with no direct sun.

FULL SHADE: This term often appears on plants that are mainly intended for outdoors. With indoor plants, full shade is equivalent to bright or medium indirect light.

PARTIAL SHADE OR MIXED SHADE: Usually used for outdoor plants. The indoor equivalent would be a mix of direct sun and bright indirect light.

Measuring Light

The human eye is a bad measuring device, since our eyes automatically adjust to different levels of light. A few tricks and tools can help us gauge the light in our homes so we can tell how much light we can offer our plants.

A shadow test will give you a general idea of whether your light is direct or indirect (filtered or diffused). Simply place your hand about a foot away from where you want to put a plant—does your hand create a defined shadow? You likely have direct sun there. Do you see a softer, less defined shadow? You have indirect light.

A light meter (or light meter app) can be a helpful tool to give you a good idea of the range of light you have available. Look for one that gives you measurements in a unit called "foot-candles" (an antiquated term that means how much light hits an area of one square foot when an even light source is in the center of it) rather than just lumens. Separate light meter devices will vary in cost and features, but if you're only using a light meter for plants, it's not worth spending a ton of money on one. Read reviews and look for one that's well-rated, can provide foot-candle measurements, and doesn't cost too much. Keep in mind that light meters are designed for photography, not plant care, so it's not the perfect tool, but it helps. For apps, there are many available, but most of the reviews you see are from photographers, whose needs are a bit different from those of us using the app for plants. The important thing to look for is accuracy and consistency in measurements. I use an inexpensive, easy-to-use app called *Light Meter* (though, at this time, it's for iPhones only. For Android users, *Lux Meter* is well-rated).

This chart gives you a reference for what type of light you have based on the foot-candle measurement from a light meter in an indoor space.

TYPE OF LIGHT PLANT NEEDS	FOOT-CANDLE MEASUREMENT
Direct Sun	1000+
Bright Indirect	600–1000
Medium	200–600
Low	50–200

HUMIDITY

Humidity is often a missing ingredient in the environment we expect plants to live in. Many houseplants are tropical plants and live in humid environments naturally—think about the feeling of walking into a greenhouse at a botanical garden or stepping off a plane in a tropical place. While some plants will do fine without humidity, even in dry climates, most will thrive best with some added humidity, and certain plants *must* have humidity in order to stay happy and healthy. (As an added bonus, humidity is also beneficial to humans—it can help with respiratory health and dry skin.) Understanding how to add humidity to your environment will give you a wider range of options when choosing which plants to bring home.

MISCONCEPTION: *Humidity is the same thing as heat.*
TRUTH: While many people just think of heat when they hear "humidity," this term actually refers to water vapor in the air. Both warm and cool climates can be humid, though warm air can hold more water vapor than cool air.

MISCONCEPTION: *Misting occasionally creates humidity.*
TRUTH: Misting can create temporary humidity, but for lasting water vapor in the air, this is not enough. Misting can also cause problems with leaves.

MISCONCEPTION: *Humidity is hard to provide.*
TRUTH: Some plants do need an extreme level of humidity, which can be a challenge to provide, but for many plants that benefit from humidity, something as simple as a pebble tray (also called a humidity tray) or a small humidifier can be enough.

MISCONCEPTION: *Humidity will ruin my home and furniture.*
TRUTH: Pebble trays or small humidifiers rarely cause damage if they're properly used and there is good air flow in the room. Be sure to aim the flow of vapor so that it's not directly hitting furniture. A fan will keep air moving and help prevent mold or damage to wood.

Three Simple Ways to Add Humidity

1. NATURALLY HUMID ROOM

A room that already has humidity (from a shower, sink, or dishwasher, for example), or an outdoor space in a humid climate is a good place for plants that need a little humidity boost. Bathrooms with a shower or bathtub that's used regularly will have more humidity than a kitchen.

2. PEBBLE TRAY

A pebble tray is simple to make and is an easy way to provide humidity for a single plant or a small grouping of plants. The water in the tray evaporates around the plant, providing the plant a personal humidity "bubble."

3. HUMIDIFIER

A humidifier is the most effective way to add moisture to the air, especially if you have a lot of plants that need humidity.

WHICH TYPE OF HUMIDIFIER SHOULD YOU USE?

- Cool vs. warm: Warm-mist humidifiers heat the water until a warm vapor is created. Evaporative cool-mist humidifiers produce vapor through the use of a fan blowing water across a filter or wick to create a fine mist. Ultrasonic cool-mist units use a nebulizer that breaks water droplets into a mist. Warm-mist humidifiers aren't as effective for large spaces but do work well in a small room or for a small group of plants. Cool-mist types are better for larger spaces. Evaporative models will need to have the filter cleaned or replaced regularly, while ultrasonic models don't have a filter but should be cleaned very often to minimize the risk of vaporizing bacteria that can form inside. All types need to be cleaned at least occasionally to remove hard water buildup, although using distilled water in your humidifier can help minimize such buildup.

- Run time: The size of the humidifier will determine its run time (how long it can run before needing to be refilled). If you plan to run the humidifier all day or don't want to have to refill it frequently, choose a model with a longer run time. Average tank sizes range from ¼ of a gallon to 5 gallons. A 1- to 2-gallon tank will be sufficient for most people's needs.

- Do you have pets or kids? A cool-mist humidifier may be better for you, since warm mist units contain hot water and could cause injury if tipped over.

- Unless it's a mini-unit, place a humidifier 3–6 feet from the plants you're providing humidity for. You want to create a microclimate for the plants, not just blast them with the vapor, so don't place it too close.

- As mentioned earlier, use distilled water to minimize hard water buildup, as hard water can clog your humidifier or create a fine dust of mineral deposits as the vapor dries.

- Clean your humidifier regularly to remove mineral buildup and bacteria (using distilled water in the humidifier will buy you more time between cleanings). Avoid using harsh cleaners, which can leave a residue that vaporizes when you turn it back on. Mild dish soap or white vinegar will do the job. Models that use a filter usually need the filter either replaced or cleaned on a regular schedule.

- Many people ask if diffusers can be used as small humidifiers. Though this isn't their purpose, they will work for a very small space, like a desk, or for one or two small plants. However, avoid using essential oils in a diffuser you're using for plants, as it may cause damage to the plants.

Plant Care Quiz

The most important part of caring for plants in our homes is knowing what we can offer them. Armed with a better understanding of light and humidity, you're on your way to making better assessments of what you can offer your plants, and you're nearly ready to start choosing your plants.

Before heading out to the plant shop or nursery, make sure you have a plant plan—know where you plan to put plants and what that space offers plants. This will ensure plant care success, which builds confidence, and confidence pushes you to try new plants, learn new things, and have even more success.

The following quizzes will help guide you as you use the flowcharts to get a clear picture of what kind(s) of environment your home provides. You'll want to use this tool while you're in your home during the daytime. You may even want to check in on the areas you're assessing throughout the day, since light can change in many rooms as the day goes on. Be realistic rather than aspirational as you make your assessment—if you have a large window but never open the blinds, assess the room with the blinds closed; if you don't currently live in a humid climate or don't have a humidifier, remember that you'll need to do something right away to provide humidity if you want to bring in humidity-loving plants.

HUMIDITY QUIZ

Is your space a bathroom where the shower runs regularly?

Is your space a kitchen where you do dishes often or create steam regularly through cooking?

Do you have a humidifier you will run in your space?

Can you set up a pebble tray in your space?

If your space is outside, do you live in a humid climate?

If you answered "yes" to any of these questions,
your space can be considered humid.

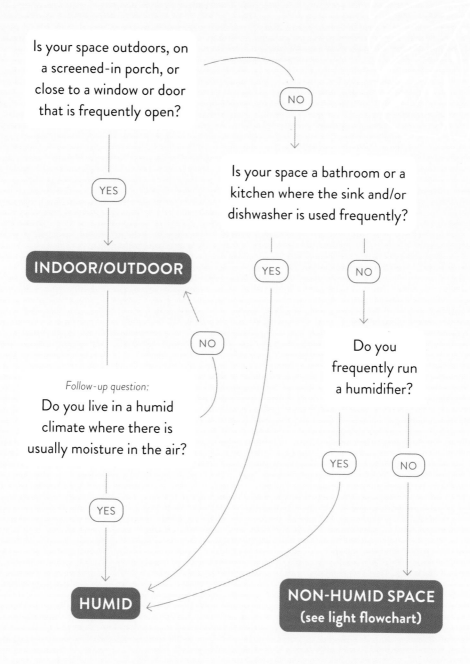

Is your space outdoors, on a screened-in porch, or close to a window or door that is frequently open?

NO

YES

INDOOR/OUTDOOR

Is your space a bathroom or a kitchen where the sink and/or dishwasher is used frequently?

YES

NO

NO

Follow-up question:

Do you live in a humid climate where there is usually moisture in the air?

Do you frequently run a humidifier?

YES

YES

NO

HUMID

NON-HUMID SPACE
(see light flowchart)

LIGHT STRENGTH QUIZ

Is your room bright for 4 or more hours a day?

Does your space get direct sun?

Is it bright enough to read or take a photo without added light?

Are your windows large and/or unobstructed?

Are your windows south facing and not covered by room-darkening curtains, blinds, etc.?

If you answered "yes" to at least two of these questions, your space can be considered bright.

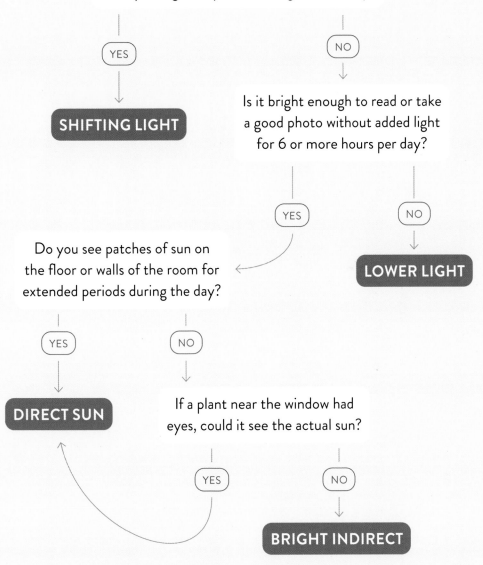

Does your light vary a lot throughout the day?

YES → **SHIFTING LIGHT**

NO → Is it bright enough to read or take a good photo without added light for 6 or more hours per day?

YES → Do you see patches of sun on the floor or walls of the room for extended periods during the day?

NO → **LOWER LIGHT**

Do you see patches of sun on the floor or walls of the room for extended periods during the day?

YES → **DIRECT SUN**

NO → If a plant near the window had eyes, could it see the actual sun?

YES → **DIRECT SUN**

NO → **BRIGHT INDIRECT**

section two

environment profiles

Now that you have an idea of what kind of environment you can provide to your plants, you're ready to dive into the details of which plants to select, how to style them, and how to care for them to keep them happy and healthy! Skip ahead to the environment profile for your space, or read through them all for a fuller understanding of where plants are happiest.

PROFILE #1:

Bright & Sunny Space

WHETHER IT'S A SUNNY LITTLE COTTAGE, an airy industrial loft, or a highly coveted apartment building graced with large windows, a brightly lit home is a dream for any plant lover. All that light certainly gives you the widest range of plant options, but it can offer unexpected challenges as well. Some plants flourish with hours to bask in the sun, while others may languish.

PROS OF A BRIGHT SPACE

+ You'll have many plant options, as most houseplants do best in bright indirect light.
+ More windows means more suitable spots to place plants.
+ Bright light fosters faster growth and bigger leaves (though this can also be a con, depending on the size of your space!).

CONS OF A BRIGHT SPACE

− East- or west-facing windows receive direct morning or afternoon sun, which can be too intense for many plants.
− More light means plants are producing more energy from the light and using more water, so they dry out faster. You'll need to water more often.
− Lots of light without added humidity can make the air drier, so humidity-loving plants will require more effort to keep happy.

TIP!

BEFORE YOU HEAD OUT TO fill those brightly lit rooms with all the plants of your dreams, you'll first want to determine if your light is direct or indirect. (See pages 4–6 for a reminder on direct vs. indirect light.)

DIRECT SUN

Placement & Styling Tips

Direct sun is great if you're dreaming of an indoor desert, but it can also be a challenge to manage. While you might be thinking, "Don't plants need lots of sun?," direct sun (or "full sun") can actually be too intense for many indoor plants, leaving you with scorched leaves and thirsty soil. If you prefer to work with your sunny space as is, you'll want to select sun-loving plants and style your room around the sun.

ARRANGE PLANTS SO THAT THEY'RE NEAR THE WINDOW but not touching the glass. Window glass gets extra hot when the sun is shining directly on it, which can scorch even the most sun-loving plants. During extreme heat waves, plants near windows may need to be pulled farther away to prevent damage from heat radiating off the glass.

DEPENDING ON THE DIRECTION YOUR WINDOWS FACE, you'll likely need to do some rearranging throughout the year as the sun's angle changes. Sun-loving plants will need at least some of that direct sun even in the winter, so reposition plants to ensure they're getting the sun they need year-round.

WITH SUCH A WIDE RANGE OF SHAPES AND SIZES, plants like cacti and succulents lend themselves well to a desert-inspired vignette. Mix taller cacti or *Euphorbia* with shorter succulents and trailing plants, like string of hearts or burro's tail, to create an instant collection. Keep things feeling cohesive by sticking to the desert palette and choosing planters in a range of shapes and shades of terra-cotta.

GOT A BIG SPACE TO FILL? Go with a statement-making oversized plant, like a bird of paradise, for major impact even with just one plant. Ready for full-on jungle vibes? Position those bigger sun-loving plants in the path of the direct sun, and place other tropical plants (which don't do well in direct sun), like *Monstera*, *Alocasia*, or *Philodendron* species, around them but away from the rays of the sun.

WHAT'S IN A NAME?

Throughout this book you'll see plants referred to by two names:

- The "botanical" or Latin name, which is the name recognized by scientists all over the world. This name includes the genus (group of plants) and species (specific plant name) of the plant, and sometimes the cultivar (variety of a plant created in cultivation).

- The "common" name. These names can vary from place to place and develop as a way to refer to plants without having to pronounce Latin words. Common names can sometimes be confusing, since they vary between places and generations, and one plant can have many common names.

- For example:
  Ceropegia woodii // String of hearts
  Botanical name // Common name

I find botanical names to be more useful, as they indicate which plants are related to each other and make any research about the plants easier to do. In the plant profiles in this chapter, I've used both the botanical and common names, in that order, because many people are more familiar with common names.

STRELITZIA NICOLAI
Giant White Bird of Paradise

LIGHT: Very bright indirect light to partial direct sun.

WATER: Keep soil evenly moist in summer months; allow to dry out slightly in cooler months. Generally, birds of paradise will tolerate drying out a bit, but if you notice leaves looking wilted, you may need to water more.

POTTING: *Strelitzia* species aren't too fussy about soil—quality all-purpose potting mix works well. Good drainage is essential.

NOTES: *Strelitzia nicolai* can grow to be 30 feet tall, so if you don't have room for a super tall plant, consider either keeping it outside if you live in a warmer climate, or opt for the more petite bird of paradise, *Strelitzia reginae*.

STYLING: *Strelitzia nicolai* can hold its own even in larger rooms, so if you're only looking for one plant to care for and have the space and light for it, look no further than these graceful giants.

BEAUCARNEA RECURVATA
Ponytail Palm

LIGHT: Very bright indirect light to partial direct sun.

WATER: *Beaucarnea* species store water in the bulbous part of their trunk (called a caudex) and have shallow roots, so allow the soil to dry out all the way to the bottom of the pot between waterings.

POTTING: Use a fast-draining potting mix, like cactus soil.

NOTES: These plants grow slowly in overall size, but their leaves can get very long quickly. They are non-toxic to pets but can make an attractive snack for cats, so if you want to keep some leaves, place it out of reach.

STYLING: Though large ponytail palms are total showstoppers, even a small one is an eye-catching addition to a credenza or multi-level plant stand.

HAWORTHIA SPECIES
Zebra Cactus

LIGHT: Very bright indirect light to partial direct sun. Too much direct sun may cause leaves on many species to turn a deep red color.
WATER: Allow soil to dry out all the way to the bottom of the pot between waterings.
SOIL: Use a fast-draining potting mix, like cactus soil.
NOTES: There are around 150 different species/varieties of *Haworthia* with varying leaf shapes and patterns, making them an easy plant to collect.
STYLING: Generally quite small in size, *Haworthia* species are a nice addition to a sunny windowsill plant gang.

CEROPEGIA WOODII
String of Hearts

LIGHT: Very bright indirect light to partial direct sun.
WATER: *Ceropegia woodii* has succulent leaves, meaning water is stored in the heart-shaped leaves, so allow the soil to dry out about ¾ of the way down the pot between waterings.
SOIL: Use a fast-draining potting mix, like cactus soil.
NOTES: String of hearts is easy to propagate from small pearl-shaped tubers that form along the vines. Simply cut a vine just above a tuber and place in soil (either back into the original plant or in a new pot).
STYLING: This plant will grow super long in a flash, so find a spot with lots of room to trail, like the top of a bookcase or hanging in a bright stairwell, so you aren't constantly needing to trim it.

DESIGN TIP! **THE ZEBRA CACTUS LOOKS** especially stunning when potted with a layer of sand or pebbles on top of the soil to highlight its distinctive white bands. Be sure soil is thoroughly dry between waterings and the planter has good drainage, as soil won't dry as quickly with sand/pebbles on top.

EUPHORBIA TRIGONA
African Milk Tree

LIGHT: Bright indirect light to partial direct sun. Too much direct sun may cause some species to change colors—often to a red, orange, or pink color.
WATER: Water when soil is dry ¾ of the way to the bottom—in colder months, this may be less than once a month. Always check before watering.
SOIL: Use a fast-draining potting mix, like cactus soil.
NOTES: *Euphorbia* species are extra toxic to pets and contain a milky latex sap that can be an irritant to skin and eyes. Use caution or wear gloves when handling.
STYLING: Use taller *Euphorbia* species to add height to a grouping of cacti and succulents, or display especially interesting specimens where they can stand out all on their own.

DESIGN TIP! *EUPHORBIA* **COME IN A WIDE** range of shapes and sizes, and many have fewer spines than cacti, so they make a good alternative if you want a desert garden look without the danger factor.

CACTI

There are more than 2,000 species in the cactus family, and while most have similar care, be sure to find care info specific to the plants you choose. These tips will give you a general idea of how to care for cacti.

LIGHT: Direct sun; some species will do fine with very bright indirect light.
WATER: Water when soil is dry all the way to the bottom—in colder months, this may be less than once a month. Always check before watering.
SOIL: Use a fast-draining potting mix, like cactus soil.
NOTES: Though the drought-tolerant nature of cacti makes them easy to care for, they do need a lot of sun, which means they should be very close to a window that gets bright light and direct sun for most of the day. If the cactus starts to change shape dramatically, this is usually an indication that the plant needs more light and is stretching out as it seeks light. Use caution when bringing a cactus into a home with kids or pets; many cacti have sharp spines that can easily get stuck in curious fingers or noses.

DESIGN TIP! COLLECTIONS OF A VARIETY OF SMALL CACTI can draw just as much attention as one large cactus, so even if you only have a small amount of direct sun, you can enjoy your own little desert garden.

Pictured: *Agave parryi, Aloe barbadensis, Crassula arborescens, Xerosicyos danguyi, Echeveria 'Black Prince,' Pachyphytum oviferum*

SUCCULENTS

A succulent is a plant that stores water in its leaves, roots, and stems. There are more than 50 plant families containing species that would qualify as a succulent, which means there are thousands of succulent plants! While many have similar care, be sure to find care info specific to the plants you choose. These tips will give you a general idea of how to care for the plants people typically think of as succulents.

LIGHT: Very bright indirect light to partial direct sun.

WATER: Water when soil is dry almost all the way to the bottom of the pot. If leaves start to look puckered/slightly shriveled, it's probably time to water. Always check soil first, especially with succulents, as even just one overwatering can doom these plants.

SOIL: Use a fast-draining potting mix, like cactus soil.

NOTES: Though most succulents can take some direct sun indoors, if you're keeping them outdoors, they usually look their best when kept in light shade.

DESIGN TIP! **SUCCULENTS OFTEN LOOK** better when grouped together as a collection. Choose a variety of species with different colors, leaf shapes, and heights for the most dynamic display.

BRIGHT INDIRECT LIGHT

Placement & Styling Tips

Bright indirect light is by far the best light for the widest range of plants. Nearly all plants will be happy in bright indirect light, even if they can tolerate less light. And even plants that like some direct sun will often be just fine in very bright indirect light.

Photo: Mila Moraga-Holz

IF YOU GET A LOT OF DIRECT SUN but want more flexibility with your plant choices, you can easily transform direct sun into indirect light with curtains or shades that filter but don't block the light. In bathrooms, window treatments like frosted glass or glass blocks let in more light than a typical bathroom window with a shade or curtain does. These treatments all provide privacy but let in plenty of bright indirect light.

MANY HOMES HAVE ONE OR TWO ROOMS with bright indirect light but other areas with less light. Place plants strategically to maximize the areas that get bright indirect light. Focus on plants that need bright light to thrive first, saving those that can tolerate less light for areas that don't get as bright.

Photo: Ron Goh

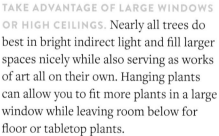

TAKE ADVANTAGE OF LARGE WINDOWS OR HIGH CEILINGS. Nearly all trees do best in bright indirect light and fill larger spaces nicely while also serving as works of art all on their own. Hanging plants can allow you to fit more plants in a large window while leaving room below for floor or tabletop plants.

DO YOU HAVE A BIG OPEN WALL IN A BRIGHT ROOM? Rather than hanging art, fill a wall with plants and create a living wall using special pockets meant for plants. Plants with vine-y growth fill in living walls quickly. Select plants with similar water requirements to make care easier.

MONSTERA DELICIOSA

Split Leaf Philodendron, Swiss Cheese
Plant, Fruit Salad Plant

LIGHT: Bright indirect light.
WATER: Allow the top 2–3 inches of soil
to dry out between waterings.
SOIL: Use rich, well-draining soil, like
all-purpose potting mix with perlite,
orchid bark, or coco coir chips mixed in.
NOTES: *Monstera deliciosa* spreads out
quickly, and as it grows, it will need something to climb, like a moss pole, redwood/
cedar stake, or trellis. With bright indirect
light and room to climb, this plant can
grow to be anywhere from 5–10 feet tall
(it can grow up to 60 feet outdoors!)
and 3–5 feet wide with interesting
leaves covered in holes and splits (called
fenestrations).
STYLING: *Monstera deliciosa*, especially
older ones, have a commanding presence
and don't need much more than an open
corner to really pack a plant punch. Pair
these with a simple planter because no
one will be paying much attention to the
planter anyway.

FICUS LYRATA
Fiddle Leaf Fig

LIGHT: Bright indirect light.
WATER: Allow soil to dry halfway down between waterings. Many *Ficus* species are sensitive to overwatering and dislike having "wet feet," so careful, consistent watering is essential.
SOIL: Use a quality all-purpose potting mix.
NOTES: Possibly the first "internet famous" plant, fiddle leaf figs are beautiful but a bit tricky to care for. To keep it happy, protect this tree from changes, like inconsistent watering and temperature fluctuations or drafts, and avoid moving it.
STYLING: If its popularity has shown us anything, it's that fiddle leaf figs are at home in many different spaces. Ready for something a little different? Try another less common *Ficus* species, like *Ficus triangularis* or *Ficus benghalensis* 'Audrey.' All three have interesting foliage and make an impact even if they're the only plant in the room.

Pictured: *Hoya compacta 'Variegata,' Hoya obovata,*
Hoya curtisii, Hoya polyneura

HOYA SPECIES

Wax Plant, Porcelain Flower

LIGHT: Bright indirect light.

WATER: Allow soil to dry out ¾ of the way down between waterings for most species. Look for slight puckering or wrinkling of leaves as a clue it's time to water.

SOIL: Use a rich, fast-draining mix, like all-purpose potting mix with orchid bark mixed in.

NOTES: Many *Hoya* species are great climbers and will climb anything you give them. Try a bamboo hoop or trellis meant for smaller plants when the plants are younger. Over time you may need to provide a larger trellis or structure. Though they can be slow growers, your patience will eventually be rewarded with some truly interesting flowers that almost look like they're made out of porcelain or sugar.

STYLING: Sculptural species of *Hoya* that trail, like the round-leafed *H. obovata*, add interest to a bookshelf.

DESIGN TIP! *HOYA* SPECIES HAVE SUCH A diverse range of leaf shapes that they make an interesting display when many species are hung in a row.

CHLOROPHYTUM COMOSUM

Spider Plant, Airplane Plant

LIGHT: Best in bright indirect light; can adapt to medium light, but variegated varieties will likely revert back to green.

WATER: Allow soil to dry out ¾ of the way down to all the way down between waterings. Leaves may turn very pale and appear extra limp when watering is overdue.

SOIL: Use a quality all-purpose potting mix.

NOTES: Spider plants are self-propagating, producing little babies on the ends of stalks that fan out from the "mother" plant. You can leave them on the mother, or remove them to root. They can even be rooted into soil while still connected to the mother plant, then removed once roots are established.

STYLING: With popularity spanning from the Victorian era to the 1970s, all the way to today, spider plants add a fun element to any room but look especially at home paired with vintage furnishings.

PILEA PEPEROMIOIDES

Friendship Plant, Chinese Money Plant, UFO Plant

LIGHT: Bright indirect light. Even light, preferably from above, will produce a plant with the most uniform shape and flattest leaves.

WATER: Allow soil to dry out ¾ of the way down between waterings. If the plant starts to droop a bit, it might be time to water. Yellow leaves, especially if they're upper leaves, usually indicate soil is too wet for how much light the plant receives. Allow soil to dry out more, or provide brighter light.

SOIL: Quality all-purpose potting mix is usually perfect. If you're having problems with the soil staying wet for too long, add orchid bark or perlite to increase drainage.

NOTES: A tradition with *Pilea peperomioides* is to share the plant with friends by removing the baby plants, which pop up near the base, when they're about 2 inches tall. The babies can be rooted, then potted into soil.

STYLING: With its very unusual shape, this plant becomes living art no matter where it's displayed. Just be sure to give it enough space for its wide span.

PEPEROMIA SPECIES

LIGHT: Bright indirect light for most species.
WATER: Allow soil to dry ⅔ of the way down, looking for slight wilting of the leaves before watering.
SOIL: Use rich, well-draining soil, like all-purpose potting mix with perlite, orchid bark, or coco coir chips mixed in.
STYLING: Smaller *Peperomia* species are well-suited for shelves and are fun to collect and group together, thanks to their varying leaf shapes, patterns, and colors.

DESIGN TIP!

THERE ARE MORE THAN 1,000 species, with leaves ranging from glossy green to textured, striped, and rose-tinted. Some, like *Peperomia scandens* or Cupid Peperomia, tend to cascade, making them a good fit for hanging planters. Pair different varieties of *Peperomia* with a range of sizes and styles of pots for added visual interest.

Pictured: *Peperomia deppeana x quadrifolia* 'Hope,'
Peperomia argyreia, *Peperomia tetragona*

Care Tips for Bright & Sunny Space

Lots of light usually encourages rapid growth and larger plants, so you'll find plants in bright rooms may need a little extra care.

WATERING: The more light, the faster a plant uses up water, especially in warm months. You always want to check the soil before watering, but in bright rooms, check it more frequently, as plants will generally need water more often.

PRUNING: More light encourages more growth, but that doesn't mean plants have to completely take over (unless you want them to). Pruning is a simple skill that you can utilize to help keep your plants at a manageable size. (See pages 153–155 to learn more about pruning.)

FERTILIZING: The more a plant grows, the more nutrients it will need to sustain that growth. Fertilizing during active growth will help a plant support new leaves, flowers, and roots. (See pages 156–158 to learn more about fertilizing.)

Photo: Sofie Vertongen

Lower Light Space

WHILE MOST OF US probably wish we had nothing but rooms with huge windows that get amazing bright light, that's usually not our reality. Take heart, though: just because you live in a compact railroad apartment, a shaded garden-level flat, or a cozy Craftsman with iconic but light-blocking deep eaves—or any other space that doesn't get a ton of light—doesn't mean you have to resign yourself to a life without plants.

PROS OF A SHADED SPACE

+ The less light a plant gets, the slower it uses the water in its soil, which means less watering is needed.
+ Plants in lower light don't grow as quickly and don't get as large as plants in bright light, which may be better for small spaces.
+ Fewer windows bright enough for plants mean plants are often clustered near those windows, which makes watering less of a chore.

CONS OF A SHADED SPACE

− Because most plants are happiest in bright light, there are fewer plant options for a space that gets lower light.
− If your dream is to see your plants get very big or full, you'll likely be disappointed.
− Small or obstructed windows means that areas even a short distance from a window get almost zero light, which limits your plant placement options.

As you start to plan the best plants for your lower light space, assess whether you have medium light or low light. There are more plants that tolerate medium light than low light, so knowing which you have is important when choosing plants. Refer back to Measuring Light (pages 8–9) to review the difference between light levels, but for quick reference:

MEDIUM LIGHT = soft shadows, 200–600 foot-candle light meter reading

LOW LIGHT = almost no shadows, 50–200 foot-candle light meter reading

MEDIUM LIGHT

Placement & Styling Tips

When light is at a premium, it is essential to make the most of what you do get. If you don't want to be limited to plants that tolerate lower light, try these tips to maximize light in your space.

REFLECT MORE LIGHT. White reflects more light, so paint your walls white (or other light colors) or hang a mirror to help direct more light to nearby plants.

UTILIZE YOUR WINDOW SPACE STRATEGICALLY. Use shelves, plant stands, and hanging plants to make space for more plants in their optimal environment.

but want a wide variety of plants, choose younger plants in smaller pots. You'll be able to fit more plants into a small space.

for the plants you want, grow lights are a good way to supplement natural light, or even add light to a room with no windows. Compared with a regular household bulb, grow lights produce more of the spectrum of light that plants need. There are many types of grow lights available, with features to meet a range of needs. Up until recently, grow lights were fairly expensive and often required complicated setups, but now it's easy to find affordable grow light bulbs that can be used in any lamp. From a style perspective, full-spectrum white LED grow lights are the least conspicuous. These white grow lights provide a balanced spectrum of light, which is effective for most houseplant needs. There are even a handful of white LED grow lights that are made to look like stylish lamps or shadow boxes (see page 56 for more on stylish grow lights).

AGLAONEMA SPECIES
Chinese Evergreen

LIGHT: Bright to medium indirect light. All-green species tolerate medium light best.

WATER: In bright light, allow the top 2–3 inches of soil to dry out between waterings. In medium light, allow soil to dry out ¾ of the way down the pot. If leaves start to droop, it's probably time to water. Yellow leaves, especially if they're upper leaves, usually indicate soil is too wet for how much light the plant receives. Allow soil to dry out more, or provide brighter light.

SOIL: Use quality all-purpose potting mix.

NOTES: Rotate plant ¼ turn each week to keep growth even. Keep leaves free of dust by cleaning them 1–2 times a month, especially if kept in lower light conditions.

STYLING: *Aglaonema* species, with their fuller, bushier shape, work well as floor plants and make a nice accompaniment to taller plants, especially trees that have less foliage at the base.

ASPLENIUM SPECIES
Bird's Nest Fern

LIGHT: Bright to medium indirect light.

WATER: Keep soil evenly moist to the touch but not soggy. Unlike most other ferns, *Asplenium* species are more forgiving of drying out a bit, but evenly moist soil is best. Avoid letting water pool at the bottom of the pot or in the saucer. Avoid getting water into the "cup" at the center of the foliage, which can cause problems for the plant.

SOIL: Potting mix with a high peat moss or coco coir content helps retain moisture without getting too soggy.

NOTES: Though more tolerant of lower humidity than other ferns, bird's nest ferns will look their best with added humidity.

STYLING: Pair these plants with a pedestal-style planter to show off the fountain-like foliage.

LOW LIGHT

Placement & Styling Tips

If you've determined that the space in which you're hoping to have plants is low light, the biggest challenge will be accepting that your options are limited. As we talked about in the light chapter, light is food for plants, so a plant in a space that doesn't get enough light will essentially starve to death. There are a handful of plants that *tolerate* low light, but this doesn't mean they do *best* in low light. Any plants included here will thrive better in more light, and most of the tips for low-light plants apply to other light levels as well.

BECAUSE YOU HAVE LIMITED options for low light, if you're looking to create a grouping of plants, try *Sansevierias*, as this genus contains over 70 species/cultivars, and nearly all of them will tolerate low light for a time. Displaying a variety of species with the same care needs together adds variety when you have limited options.

IF YOU HAVE OTHER AREAS that get brighter light, consider rotating your lower-light plants between the two spaces. This isn't ideal, as plants do best when kept in consistent conditions, but it will help the plants kept in low light to "recharge" a bit, which will help extend their lifespan. Some of our commercial clients at Folia will even have us bring two of the same plant. They'll put one in a bright light area and one in a low-light area, and then rotate them back and forth monthly. That way the space always looks the same, even with the plants rotating in and out.

WHEN YOU'RE GROUPING PLANTS in a small space, the easiest way to achieve a collected rather than cluttered look is to stick to one or two colors of planter. Whether it's glazed white pots, a collection in all different shades of pink, a group of black-and-white patterned pots, or the natural earthy tones and textures of terra-cotta, letting a color palette guide your grouping will keep your collection looking cohesive. Vary the shape and size of the pots if you'd like to avoid the matchy-matchy vibe.

ZAMIOCULCAS ZAMIIFOLIA

ZZ Plant, Zanzibar Gem

LIGHT: Any light from bright to low indirect light. Tolerates low light better than most plants. Avoid direct sun to keep leaves looking their best.

WATER: Allow soil to dry all the way to the bottom of the pot between waterings, especially if kept in low-light conditions. Good drainage is essential for ZZs, so make sure any excess water in the saucer or cachepot is emptied after watering.

SOIL: Use a fast-draining potting mix like cactus soil or all-purpose potting mix with perlite or orchid bark mixed in.

NOTES: A handful of new cultivars of ZZ plants have recently been introduced, so keep an eye out for varieties like ZZ 'Raven' with leaves that appear almost black or ZZ 'Zenzi'—a dwarf variety that stays more compact than classic ZZ plants.

STYLING: ZZs work well both solo or grouped with other plants with varying shapes. If you're working with a low-light space, larger specimens in a plant stand or sculptural planter will create an impactful plant moment.

Pictured: *S. patens, S. trifasciata* 'Black Coral,' *S. trifasciata* 'Moonshine,' *S. trifasciata* 'Hahnii,' *S. ehrenbergii*

SANSEVIERIA SPECIES
Snake Plant, Mother-in-Law's Tongue

LIGHT: Any light from bright to low indirect light. Tolerates low light better than most plants. Avoid direct sun to keep leaves looking their best.

WATER: Allow soil to dry all the way to the bottom of the pot between waterings, especially if kept in low-light conditions. Good drainage is essential for *Sansevieria* species, so make sure any excess water in the saucer or cachepot is emptied after watering.

SOIL: Use a fast-draining potting mix, like cactus soil.

NOTES: As snake plants grow, watch for signs your plant is ready for a bigger pot. Signs include roots coming out of drainage holes, bulging or deformed plastic pots, and new growth pushing up right against the side of the pot. The plant's rhizomes (thick, modified stems that grow underground and from which new offset plants form) are actually strong enough that they can break pots as they fight for more space to grow.

STYLING: Larger, more sculptural *Sansevieria* species command enough attention to display on their own. Mix more common *Sansevieria* species with unusual ones in a group to highlight the different colors, shapes, and patterns of these hearty plants.

Care Tips for Lower Light Space

Placing plants in lower light spots comes with a risk that they won't receive enough light, which will eventually be their demise. Spotting the signs your plant isn't getting enough light early can give you time to move them before it's too late. Some common signs are:

- **LEGGY, PALE GROWTH, KNOWN AS _ETIOLATION_:** As a plant attempts to seek out more light, it will stretch out its stems, its leaves and stems will lose color, and leaves will become spaced farther apart.

- **PALE LEAVES OR LOSS OF MARKINGS:** Plants in low light will often become paler, and plants that previously had strong markings or patterns on the leaves will lose them. Variegated plants (plants with multiple colors on their leaves) may begin to revert back to the all-green version (since more green means more chlorophyll and thus more opportunity to absorb more light).

- **NO NEW GROWTH OR VERY SMALL LEAVES:** Plants that are starving for light will conserve energy by slowing down growth, and any new growth will often be smaller than it should be, given the age/maturity of the plant.

If you notice these signs in your plants, it's best to move them to a spot that gets brighter light to keep them happier long-term.

If you plan to keep plants in medium or low light, here are some tips on their care:

WATERING: Because plants in low light won't be producing as much energy, they won't be able to use the water in their soil as quickly as the same plant in bright light. Low-light areas are often not as warm as bright-light areas, so less soil moisture is lost to evaporation as

well. In order to prevent root rot due to soggy soil, it is especially important to provide good drainage and ensure soil is allowed to dry to the proper level for your particular plant.

GROWTH: Plants grown in low light will not be able to grow as quickly, as large, or as full as the same plant grown in bright light. Having realistic expectations of your low-light plants will save you from worry or frustration when you aren't seeing new growth as quickly or as often.

PROFILE #3

Humid Space

FROM A PATIO OR LANAI in a steamy
locale to a plant lover's dream bathroom
(one with a window), spaces with natural
humidity are an ideal environment for
many houseplants.

If you've ever basked in the cool, damp mist coming off a waterfall or stepped out into the warm, thick air of a muggy summer day, you've encountered humidity. Naturally humid environments exist in places as diverse as the cool rainforests of the Pacific Northwest to the steamy jungles of Southeast Asia, and each hosts a wide range of plant species. Plants are remarkably adaptable, as the houseplants living in our homes—an environment far different than their natural habitats—illustrate. While some plants can easily adjust to life in our homes, others need at least a little bit of what they'd have in the wild in order to thrive indoors. If you live in a humid climate or have a naturally humid room with enough light for plants, you have an advantage over those of us who don't.

PROS OF A HUMID SPACE

+ Since many houseplants are tropical, a humid home environment will be their happy place, as it more closely mimics their native setting.
+ Plants that are more difficult to keep in dry climates or dry rooms will thrive in humidity, so you can keep tricky plants with less effort.
+ Since humid spaces have more moisture in the air, you normally won't need to water as often.

CONS OF A HUMID SPACE

− If not managed properly, the higher humidity levels you might need for some plants can create problems for wood furniture, doors, walls, and carpeting. You'll need good air flow, usually through fans, to keep mildew and other damage at bay. See pages 86–87 for more tips on dealing with problems humidity may cause.
− Higher humidity levels, especially without enough air flow, can cause more soil fungi (mold) to bloom. While these types of fungi won't harm the plant, they can be a nuisance and don't look as attractive as bare soil.
− Plants native to arid climates, like cacti, may not do as well in humid spaces.

Placement & Styling Tips

BATHROOMS

If you're lucky enough to have a bathroom with a window, you have a room that's already well suited for humidity-loving plants, as long as there's enough light for the type of plant you choose. Even in a smaller bathroom, there are some simple ways to make room for plants:

USE VERTICAL SPACE BY INSTALLING SHELVES or turning an over-the-toilet shelving unit into a plant shelf!

HANG PLANTS TO TURN OVERHEAD SPACE into a floating jungle. If you have high ceilings above the shower or tub, install a second shower curtain rod higher up to really max out the available plant space. If you're a renter and concerned about installing hardware, quality tension rods will usually work for this purpose. As an added bonus, when you water, you won't even need to move the plants while they drip dry.

IF YOUR SHOWER RECEIVES NATURAL LIGHT, use waterproof adhesive-backed hooks (like 3M brand Command Hooks) to hang mounted plants on the tile. Keeping plants at one end of the shower, away from the showerhead, will protect them from soapy splashes.

IF YOU'RE USING THE CACHEPOT system (a plastic grower pot placed inside a decorative pot—see page 123 for more info), choose a decorative pot that's slightly deeper than you need. Place pebbles in the bottom of the pot, and you have a pebble tray hidden inside the pot! Just don't forget to make sure the water doesn't touch the bottom of the grower pot, and change the water at least once a week to fend off pests and stagnation.

NO WINDOW IN YOUR BATHROOM? Another commonly semi-humid spot in any home is near the kitchen sink (as long as you use the sink regularly). Create a mini windowsill jungle with an assortment of smaller humidity lovers to admire while you do dishes or prep meals. If you want to be doubly sure you have enough humidity, use a long rectangular tray as a pebble tray and arrange all the plants on it. You don't have to sacrifice style to create humidity with a pebble tray.

GROUP SMALLER PLANTS TOGETHER and display on a decorative waterproof-tray-turned-pebble-tray for added humidity. Choose plants with a variety of shapes, sizes, and heights to add interest.

PORCHES AND PATIOS

If you're not ready to go full-on outdoor jungle on your humid porch or patio, start with just a few strategic choices that make a big impact. This is also a wise approach in climates where temperatures drop in the winter if you don't have a lot of inside space for plants, since plants will need to be brought indoors (see page 96 for more info about over-wintering plants).

IF STRONG WINDS OR HEAVY RAINS are a concern, a few easy steps will keep plants more secure:

- Place taller plants closer to walls or other sheltered areas, and plant in heavier pots to add stability and prevent tipping over.

- Place hanging plants in moss-lined wire baskets for faster draining and less risk or breakage than plastic. Use chain and heavy-duty carabiners to hang the baskets and keep them more secure on windy days.

EVEN A SINGLE PALM WILL ADD LUSH tropical vibes to an open corner and add privacy.

Photo: Dabito

HANG SUPER-FULL PLANTS, LIKE BOSTON FERNS, on a covered porch to create a green canopy and help buffer noise from the street or neighboring houses.

ALOCASIA SANDERIANA X WATSONIANA

Alocasia amazonica, African Mask

LIGHT: Bright indirect light.

WATER: Keep soil evenly moist to the touch but not soggy. Provide added humidity through a pebble tray or humidifier.

SOIL: Potting mix with a high peat moss or coco coir content helps retain moisture without getting too soggy.

NOTES: *Alocasia* species, like this hybrid of two other species, can go dormant if temperatures drop too low. If the leaves on yours start to die off in cooler months, it's likely just going dormant. If all the leaves die, reduce watering, but don't allow the soil to dry out all the way. In the warmer months, when growth starts to pop back up, begin watering as usual.

STYLING: *Alocasia amazonica* has such interesting leaves that it makes the perfect addition to an otherwise common grouping of plants. Or place it on a plant stand to really draw attention to it.

CALATHEA SPECIES
Prayer Plant

LIGHT: Bright to medium indirect light.

WATER: Keep soil evenly moist to the touch but not soggy. *Calathea* species can be especially sensitive to cold water and hard water, so room temperature distilled water will keep them happier (at a minimum, ensure tap water is room temperature before using).

SOIL: Potting mix with a high peat moss or coco coir content helps retain moisture without getting too soggy.

NOTES: Added humidity is essential to keep *Calathea* species looking their best. Brown edges are usually a sign of too-low humidity levels. *Calathea* species are referred to as "prayer plants" because at night (or when light levels drop), their leaves slowly fold inward, then open back up during the day (or when light levels go back up). This is called nyctinastic movement and is an interesting phenomenon related to the circadian rhythm of certain plants.

STYLING: *Calathea* species have patterned, showy leaves, and shine when displayed solo or grouped with other plants with less patterned leaves. Pair them with simple pots that don't compete with their interesting colors and patterns.

FITTONIA ALBIVENIS
Nerve Plant

LIGHT: Bright indirect light.
WATER: Keep soil evenly moist to the touch but not soggy.
SOIL: Potting mix with a high peat moss or coco coir content helps retain moisture without getting too soggy.
NOTES: Nerve plants are stunning but can be a little tricky to grow. Keep in high humidity and water carefully. If the whole plant wilts, a thorough watering will often revive it.
STYLING: Because of their particular needs, a terrarium is the best place to display a *Fittonia* species. Choose a glass vessel that suits your style, and the nerve plant's high-contrast leaves will make your terrarium an eye-catching focal point of any table.

HOWEA FORSTERIANA
Kentia Palm

LIGHT: Bright indirect light indoors, partial to full shade outdoors.

WATER: Allow the top 2–3 inches of soil to dry out between waterings. Humidity will keep fronds looking their best.

SOIL: Use a fast-draining potting mix, like cactus soil.

NOTES: Kentia palms can tolerate medium light indoors but won't grow as vigorously. Given some time outdoors, they may be able to take some morning direct sun, but if leaves exhibit burnt spots, move to full shade.

STYLING: Taller Kentia palms are an easy way to break up a large space with foliage or add privacy when placed in front of a window or in an outdoor living space.

Pictured: *Nephrolepsis exaltata, Pellaea rotundifolia, Pteris cretica, Arachniodes simplicior* 'Variegata'

FERNS

LIGHT: Bright to medium indirect light.
WATER: Keep soil evenly moist to the touch but not soggy.
SOIL: Potting mix with a high peat moss or coco coir content helps retain moisture without getting too soggy.
NOTES: Though care may vary slightly between different kinds of ferns, nearly all will be happiest in a humid environment. Over time, as a well-cared-for fern matures, brown dots will form on the underside of leaves—these are the spores ferns use to reproduce and are a normal sign of a healthy plant.
STYLING: Because of the range of leaf shapes and varied shades of green ferns encompass, they look amazing displayed in a collection. Very full species, like *Nephrolepsis exaltata*, a.k.a. Boston ferns, are great for filling an open corner when hung or placed on a plant stand.

MESIC *TILLANDSIA* SPECIES
Air Plants

LIGHT: Bright indirect light.

WATER: Water by completely submerging the plant every 10 days for 2 hours. Remove from water and allow to dry upside down for at least 4 hours or until completely dry. (I stick mine in the dish drainer to dry.) This will prevent water from getting trapped in the base of the plant, which can lead to rot.

SOIL: *Tillandsia* species are epiphytes, meaning they do not grow naturally in soil or need it to survive. Because they don't live in soil, adding fertilizer to their soaking water is the only way they can get nutrients indoors. (See Fertilizing, a.k.a. Feeding, pages 156–158.)

NOTES: Mesic *Tillandsia* species are a group of air plants native to humid climates. They can be identified by their smooth leaves (rather than the fuzzy leaves seen on xeric *Tillandsia* species, which are native to more arid climates) and are usually green, or sometimes shades of red, pink, or purple, but never silver or white.

STYLING: Air plants are fun because they're not bound by being in a pot! Just make sure wherever you place yours provides bright indirect light and isn't too inconvenient to reach, since you'll need to take it down to water.

DESIGN TIP!

AIR PLANTS INVITE YOU TO be creative. Try them on the wall in a stylish hanger made for air plants or sitting in a special dish as part of a coffee table or bookshelf vignette. Some people even fashion simple hangers for them out of wire and hang them in groups where they almost appear to float.

MARANTA SPECIES
Herringbone Plant, Prayer Plant

LIGHT: Bright to medium indirect light.

WATER: Keep soil evenly moist to the touch but not soggy. *Maranta* species can be especially sensitive to cold water and hard water, so room temperature distilled water will keep yours happiest (at a minimum, ensure tap water is room temperature before using).

SOIL: Potting mix with a high peat moss or coco coir content helps retain moisture without getting too soggy.

NOTES: Like their cousin *Calathea*, added humidity is essential to keep *Maranta* species looking their best. Brown edges are usually a sign of too-low humidity levels. And just like *Calathea*, *Maranta* species are also referred to as "prayer plants" because of the way their leaves fold up at night.

STYLING: *Maranta* species have very striking leaves with a high-contrast pattern and bright pink or neon green veins. Their graceful draping shape makes them perfect for a shelf, especially one at eye level where that striking foliage can be easily appreciated.

SAINTPAULIA IONANTHA
African Violet

LIGHT: Bright indirect light.
WATER: Allow soil to dry out to just below the surface. As soon as the surface feels dry to the touch, water. The leaves can be easily damaged by water spots, so bottom watering is safest. Ensure the plant is removed as soon as the soil has been saturated.
SOIL: Potting mix with a high peat moss or coco coir content mixed with perlite helps retain moisture without getting too soggy.
NOTES: African violets don't absolutely need humidity to survive but will thrive best with some added humidity. A pebble tray is usually adequate.
STYLING: With blooms in an array of colors from pinks and reds to blues and purples, African violets are one of the easiest flowering indoor plants, and they add a pop of color to shelves or window-sills in a way few other indoor plants can.

Care Tips for Humid Space

Are you drawn to all these humidity-loving plants but you live in a dry climate? It's possible to create a humid zone for plants in any climate! Here are a few ways to do that:

- Create pebble trays (page 12) and group humidity-loving plants together on them.

- A good humidifier can turn any space into a place humidity-loving plants can be happy. Refer back to pages 15–16 to learn more about humidifiers and how to use them.

- Misting is not a very effective way to provide continuous humidity, but it can work well for plants like ferns or air plants, especially between waterings.

DEALING WITH HUMIDITY PROBLEMS

As I mentioned earlier, high humidity levels can cause problems for both plants and our homes if not managed properly. While many of these problems are only an issue in spaces with extreme humidity, if you are experiencing them, a few simple measures can prevent or limit them.

- **AIRFLOW:** Indoors, moving air is one of the easiest ways to prevent mildew on walls and furniture, and soil fungus or more serious fungal diseases on plants. Running a fan or opening a window in a bathroom, or any room you're adding humidity to, will minimize the risks high humidity can bring.

- **PRUNING:** Outdoors in humid climates, some plants may get more airflow than others, based on their placement in relation to other plants. If you're finding plants affected by fungal diseases outside, you may need to prune the surrounding plants to increase airflow to the diseased plants. Indoors, plants with dense foliage close to the soil surface may also need to be pruned to manage fungal issues. (See pages 153–155 for more on pruning.)

- **CLEANING:** If you're using a humidifier to create a humid space, be sure to clean it regularly, based on the manufacturer's suggestions. If your humidifier has mildew growing in it, you're actually vaporizing the mildew spores whenever you run it.

(Gross, I know.) You may also need to wipe your walls down regularly with a sanitizing cleaner to eliminate any mildew that has formed.

- **POSITIONING:** Make sure your humidifier is placed so that the water vapor is falling as evenly as possible onto the plants. The exact placement will vary depending on the type of humidifier you have, but it's easy to observe the vapor to see how it's falling. It's helpful to have a model where you can adjust the aim of the vapor spout as needed. Avoid placing humidifiers too close to the wall, and make sure a single leaf isn't blocking the flow of the vapor, causing water to drip into just one area as it collects on the leaf.

Indoor-Outdoor Space

IF YOU HAVE OUTDOOR SPACE, whether a tiny balcony or fire escape, or a full patio or wraparound porch, you know how amazing it can be to have room to expand your houseplant jungle (and your living space too!) in the warm months. Plants that you'd never be able to keep happy inside thrive outside. Just by opening your doors, your inside plants can blend into your outside plants, magically making your space feel bigger. Indoor-outdoor space is a happy place for both plants and people.

PROS OF AN INDOOR-OUTDOOR SPACE

+ Utilizing outdoor space gives you even more room for plants.
+ Because the light outside is more ample than inside, it opens up the possibility to keep a wider variety of plants.
+ Not being bound by ceiling height will give you space for larger plants that you might not have room for indoors.
+ Thanks to predatory insects, who eat many of the plant pests, problem pests can often be managed more easily outdoors.

CONS OF AN INDOOR-OUTDOOR SPACE

– Keeping plants outdoors is easiest in temperate climates. In cooler climates you'll need to bring plants inside for the winter months, which may be a challenge if you have limited space or low light indoors.
– There are more conditions to manage outdoors, like direct sun, outdoor pests and diseases, wind, and drastic temperature changes.

Placement & Styling Tips

Plants make it easy to blur the lines between the indoors and outdoors, which can be especially helpful for smaller homes or apartments to make the space feel bigger. Try placing plants both inside and outside near patio or balcony doors to easily connect the spaces visually. Hanging plants above patio doors add extra lushness as you transition from indoors to out.

If your neighbors are a little too close for comfort, create more privacy and beautify your space at the same time by placing plants strategically. Line up taller plants, like trees or larger tropical plants, or hang long, full plants to create a living screen that can be moved around if needed.

CONDITIONS WILL, OF COURSE, BE much more extreme outdoors, with larger temperature swings and variations in wind and sun, so keep that in mind when choosing where to place plants. Arrange plants based on light needs, selecting sun-loving plants for spots that get direct sun (especially afternoon direct sun, as it is more intense than morning direct sun), and then place tender plants in the areas more hidden from the sun. Remember that shade outside = bright indirect light inside, so shade is your friend when keeping most plants outside. No shade in your space? Use shade cloth or a shade sail to create shade, or focus on those sun-loving plants instead.

Photo: Sofie Vertongen

IF YOUR ONLY OUTSIDE SPACE is a small balcony or fire escape, maximize your space with vertical planting. Wall pockets, like the ones made by WallyGro, make creating a living wall easy—just follow their simple installation instructions and fill with the right plants for the light you get. No wall space? Make use of balcony railings by employing planter boxes made for railings (available at most garden centers), or train plants placed on the floor to grow up the railings instead.

ALOCASIA 'REGAL SHIELDS'
Elephant Ear, Giant Taro

LIGHT: Bright indirect light indoors, full shade if kept outdoors.

WATER: Allow the top 2 inches of soil to dry out between waterings—*Alocasia* species will not tolerate drying out further than that. These do best with humidity, so if you're keeping yours outside in a dry climate, you'll want to mist it with a hose or large spray bottle once or twice a day. It will likely need water every day outdoors in a dry climate as well.

SOIL: Use rich, well-draining soil, like all-purpose potting mix with perlite, orchid bark, or coco coir chips mixed in.

NOTES: Elephant ear plants can grow to be massive, especially when kept outside. This tropical plant will be happiest when temperatures are above 55–60° F, so keep an eye on the temps in the fall and bring inside once they start to drop below 55° F.

STYLING: If you're blessed with a water feature in your outdoor space, your elephant ear will be happiest near it. Combined with proper watering, the humidity from ponds or fountains will help prevent brown edges and droopy leaves and bring an instant jungle vibe to your water feature.

THAUMATOPHYLLUM BIPINNATIFIDUM

Lacy Tree Philodendron

LIGHT: Bright indirect light indoors, partial to full shade if kept outdoors.

WATER: Allow the top 2–3 inches of soil to dry out between waterings.

SOIL: Use rich, well-draining soil, like all-purpose potting mix with perlite, orchid bark, or coco coir chips mixed in.

NOTES: You may know this plant as a philodendron, but recent genetic evidence caused a reclassification to the genus *Thaumatophyllum*. These plants form a woody, tree-like stem marked by white scars from old leaves shed over time. Many plants in this genus grow to be quite large, especially when grown outdoors.

STYLING: Because of their potentially massive size, *T. bipinnatifidum* are perfect for filling a corner of a larger patio or porch, as they will spread out wide when given enough room.

DESIGN TIP!

IF THE WILD NATURE OF *T. bipinnatifidum* gets to be overwhelming, utilize stakes, a moss pole, or a trellis to control the shape by gently directing leaves away from walkways, etc.

HANGING SUCCULENT VARIETIES

LIGHT: Very bright indirect light to partial direct sun indoors, partial to full shade when kept outdoors.

WATER: Water when soil is dry almost all the way to the bottom of the pot. If leaves start to look puckered/slightly shriveled, it's probably time to water. Always check soil first, especially with succulents, as even just one overwatering can doom these plants.

SOIL: Use a fast-draining potting mix, like cactus soil.

NOTES: Though succulents can take some direct sun indoors, if you're keeping them outdoors, they usually look their best when kept in light shade.

STYLING: Hanging succulents make excellent "curtains" on a balcony or patio. Whether hanging or planted in a railing planter, they will fill in empty space nicely and can take some direct sun, so they're perfect for any outdoor space that gets some full sun. Just make sure they're not in full sun all day long, as they will start to look sad.

Pictured: *Senecio radicans, Senecio rowleyanus*

Pictured: *Mammilaria bocasana, Cleistocactus strausii, Pilosocereus pachycladus, Opuntia engelmannii*

CACTI

LIGHT: Direct sun indoors. Direct sun to partial shade outdoors. Some will do fine with very bright indirect light indoors.

WATER: Water when soil is dry all the way to the bottom—in colder months, this may be less than once a month. Always check before watering.

SOIL: Use a fast-draining potting mix, like cactus soil.

NOTES: Though the drought-tolerant nature of cacti makes them easy to care for, they do need a lot of sun, which means they should be very close to a window that gets bright light and direct sun for most of the day. If the cactus starts to change shape dramatically, this is usually an indication that the plant needs more light and is stretching out as it seeks light. Use caution when bringing a cactus into a home with kids or pets, as many cacti have sharp spines that can easily get stuck in curious fingers or noses.

STYLING: Because of the huge variety of shapes, sizes, and colors in the cactus family, they make a spectacular grouping outdoors, even when displayed simply in terra-cotta pots and placed on cinder-block shelves.

Care Tips For Indoor-Outdoor Space

Watering outdoor plants, even if they're in containers, will take some getting used to if you've only ever had indoor plants. Because it is brighter and windier outdoors with more temperature variation, soil dries out much faster, and plants go through water faster too. Don't be surprised if you're watering the same plant two to four times as frequently each week when it's placed outside vs. inside.

If you keep houseplants outside and live in a non-temperate climate where temperatures regularly drop below 50° F, you'll need to have a plan in place for caring for those plants over the winter. Knowing what you have room for inside *before* you purchase plants for outside will ensure you're prepared when they have to come in from the cold. Keep in mind available window space, since light levels in general are also lower during the shorter days of winter.

Transitioning plants from outside to inside is a process. You'll want to give each plant a good checkup and cleaning to ensure no pests hitch a ride into your house. There will also be an adjustment period for the plants as they reacclimate to an indoor environment and lower light levels. Try to keep them in the brightest light you can for the first week, then slowly move them if you need to, keeping their optimal light needs in mind. Don't be surprised if they drop leaves, look droopy, or otherwise struggle initially when transitioning to an indoor space.

Shifting Light

AFTER READING THROUGH all these environment profiles, you may be thinking, "my space is a mix of these," and you're not alone. Many of our homes have rooms where the time of day affects the plant environment. Perhaps you live in a house with large windows at one end of an open-layout living room that gets great light in the morning, but that light drops off throughout the day. Or you call a corner apartment home where it's dim until the end of the day, when the sun blasts in. Or maybe you just live in a locale that's bright and sunny in the summer but overcast and cloudy all winter. Spaces with shifting light are very common, and now that you have an understanding of each type of light, you'll be ready to make the most of any light you get.

Photo: Christel Robleto

PROS OF SHIFTING LIGHT

+ Working with a variety of light will open up the possibility of expanding your plant collection to include a broader variety with a range of needs, which will build your plant knowledge.

CONS OF SHIFTING LIGHT

- Creating a care routine for plants with different needs can be more challenging since they won't all need water or fertilizer at the same time.
- Your plants may be spread around your home, which means more effort when watering and more risk of overlooking a plant in need.

Placement & Styling Tips

EVEN IF YOUR SPACE IS A SINGLE ROOM WITH SHIFTING LIGHT, you can still set up mini climate zones in that room to accommodate a variety of plants. Do you have a thing for humidity-loving plants? Group ferns, *Calatheas*, and other plants that thrive in humidity (see pages 76–85) together on one large pebble tray or arrange them around a small humidifier to create a humidity "bubble" in an otherwise dry room. Do you have a small window that gets direct sun? Gather desert plants, like cacti and succulents, on the windowsill to bask in the glow.

**WHEN YOU'RE WORK-
ING WITH SHIFTING
LIGHT,** you'll find it
easiest to group plants
by their light needs.
Determine what light
levels you're working
with for each area you
want plants in, then
group together the
plants best matched to
that light.

LIGHT CAN BE MANAGED with window treatments. If your room faces north or only gets morning light, make the most of it by keeping the blinds open (use the plants themselves to create privacy, if that's a concern). Too much direct sun in the afternoon? Hang a white, semi-sheer curtain to transform that sun into bright indirect light.

WHEN LIGHT SHIFTS SEASONALLY, there are adjustments you can make to help plants weather the changes:

- Rearrange your plants: Move them closer to the windows in the darker months and farther away in brighter months. In darker winter months, you may find you have to place more plants near the windows than you'd have there in the brighter months.

- Don't be afraid to use supplemental grow lights. There are many versions available today that seamlessly blend into most decor styles. Personally, I'm a fan of Modern Sprout's line of LED grow lights, which range from a simple and stylish bar that mounts under a shelf or can be suspended from the ceiling to frames that can be mounted on the wall and come in a handful of fun colors. The pendant grow lamps made by Soltech Solutions also have a simple, modern design, and utilize LEDs as well.

- Focus on plants that can adapt to or tolerate medium and low light so that you don't have to worry as much when the seasons change. All the plants listed in this chapter and the Lower Light chapter will be good options for you.

PHILODENDRON HEDERACEUM
Heartleaf Philodendron

LIGHT: Bright to medium indirect light.
WATER: Allow the top 1–2 inches of soil to dry out between waterings. Leaves will begin to look soft and droopy when the plant needs water.
SOIL: Use rich, well-draining soil, like all-purpose potting mix with perlite, orchid bark, or coco coir chips mixed in.
NOTES: This *Philodendron* species is usually sold as a hanging plant, but if given something to climb, the leaves will mature and take on a new shape.
STYLING: These are fast growers and grow to be quite full (especially in bright light), so they make excellent privacy plants for hanging in windows with no window treatments.

SCINDAPSUS PICTUS 'EXOTICA'

Satin Pothos

LIGHT: Bright to medium indirect light.
WATER: Allow the top 1–2 inches of soil to dry out between waterings. Leaves will begin to curl under when the plant needs water.
SOIL: Use rich, well-draining soil, like all-purpose potting mix with perlite, orchid bark, or coco coir chips mixed in.
NOTES: Though the common name for *S. pictus* 'Exotica' is "satin pothos," this plant is in a separate genus from the other plants referred to as *pothos* (*Epipremnum*). There are other varieties of *Scindapsus pictus*, but 'Exotica' is the most tolerant of medium light.
STYLING: The silvery, sage green colors of *Scindapsus pictus* make them stand out among other green hanging plants. Pair them with dark green plants, like *Ficus elastica* 'Burgundy' or *Alocasia amazonica*, to really show off the silvery tones.

PEPEROMIA OBTUSIFOLIA
Baby Rubber Plant, Pepperface Plant

LIGHT: Bright to medium indirect light.
WATER: Allow the soil to dry out ⅔ of
the way down between waterings.
SOIL: Use quality all-purpose potting mix.
NOTES: *Peperomia obtusifolia* is the
heartiest of the *Peperomia* species, so if
you've struggled with other species, like
P. caperata, take heart, because these are
much easier to care for.
STYLING: As they grow, pepperface
plants take on a graceful shape that
drapes downward, making them a perfect
addition to a bookcase or wall shelf.

DESIGN TIP! **BECAUSE *PEPEROMIA***
species are non-toxic to animals,
they are a great option for homes
with pets. If your furry friend is snacking on
them a little too much, you may want to group
your *Peperomia* on higher shelves to keep them
looking their best.

DRACAENA WARNECKII
Dragon Tree

LIGHT: Bright to medium indirect light.
WATER: Allow the soil to dry out ⅔ of the way down between waterings.
SOIL: Use a fast-draining potting mix, like cactus soil or even cactus soil with small lava rocks mixed in (many *Dracaena* species will come potted in mostly lava rock).
NOTES: If the tips of the leaves on your *D. warneckii* are turning brown, you may be letting the soil get too dry, or your water may be too high in mineral salts or chlorine. Using distilled or rainwater can help solve this (see page 147 for more on using distilled water).
STYLING: *Dracaena* species is one of the only trees that tolerate medium light well, so for height in spaces that aren't as bright, this is your tree.

DESIGN TIP! **THE LEAVES ON *DRACAENA*** species range from thin to wide and come in an array of colors, and the older plants often take on interesting shapes that make them a perfect statement plant when styled alone.

EPIPREMNUM AUREUM
Jade Pothos, Devil's Ivy

LIGHT: Bright to medium indirect light.
WATER: Allow the top 1–2 inches of soil to dry out between waterings. Leaves will begin to look soft and droopy when the plant needs water.
SOIL: Use rich, well-draining soil, like all-purpose potting mix with perlite, orchid bark, or coco coir chips mixed in.
NOTES: Most *Epipremnum* (usually referred to by their common name "pothos") are typically sold as hanging plants, but if given something to climb, the leaves will grow very large, and as they mature will take on a new shape and even develop fenestrations (the technical term for splits and holes in plant leaves), similar to many *Monstera* species. Though there are a number of common varieties of *Epipremnum aureum*, the all-green 'Jade' or green with marbled yellow 'Golden' tolerate medium light best.
STYLING: *Epipremnum aureum* are also fast growers and climb easily, so they can be trained to climb anything from stair rails to the frame of a bookcase, or even fishing line/twine stretched across the wall in the pattern you'd like the plant to grow.

Care Tips for Shifting Light

Caring for plants with different watering needs can be a little tricky, but creating a routine will make it easier. Divide up plants by their watering needs and assign a different check-in day to each group. Depending on the kind of plants you have, some groups will only need water every other week, others once a week, some possibly every few days. If you have a larger number of plants, this system makes watering days less overwhelming, since you won't have to check every single plant all at once.

If your plants are spread around the house in order to take advantage of each light type, this is the system I like to use to make watering days more efficient:

- For plants you water at the sink/tub, start by gathering all the plants in smaller pots and watering those at the sink. If you have a number of small plants, you may find it easiest to bottom water (see page 151). Next, move on to medium-large plants and water them in the tub.

- Round up the hanging plants and water them at the tub as well. Hang on the shower curtain rod/shower wall to drip dry (lay down towels if drips are a concern).

- While the sink/tub plants are draining (or bottom watering), water the plants that are too big to move or that you prefer to water in place.

plant care essentials

Bringing plants into our homes opens a door to a life of learning if we're committed to helping them thrive. Even failures teach us something. In fact, some of the best lessons can come through killing plants. While failure plays an important role, it can also be discouraging and trick us into thinking, "I just have a black thumb; maybe plants aren't for me." Having a foundation of plant care know-how will help you fail less, feel more empowered, and see happier, healthier, plants.

Basics of Plant Care

MATCHING THE RIGHT PLANTS to your space will give you the best chance for success, but learning how to provide proper care is essential to keeping your plants happy long-term. From selecting healthy plants to potting, watering, pruning, and everything in between, these care basics are the building blocks to seeing your plants thrive.

A FEW THINGS TO KEEP IN MIND

- Plants are living things, not props. Leaves may fall off, grow to be oddly shaped, or turn brown or yellow on the edges. Sometimes these are signs of a health problem, other times these are just part of the natural life cycle of a plant. Just because a plant doesn't look perfect doesn't mean there's something "wrong" with it.

- Plant care involves lots of trial and error. If something doesn't work, take note of what happened, do some research to try to find out why, and try again!

- Plant care methods can be personal, as they are derived from what we've learned from family, friends, teachers, and other sources, and there are often several ways to achieve the same goal with plants. There is also a lot of information about plant care out there, so use wisdom in choosing the source of your care info. Look to plant specialists rather than decorating generalists or trend-hopping websites. Seek out people who actually care for and work with a wide variety of plants in an indoor setting. Plant shops that specialize in indoor plants are often a good place to start. Many botanical gardens have an indoor plant care specialist who may offer classes or talks, and you may be able to find plant clubs or garden clubs in your area that are also helpful.

CHOOSING YOUR PLANT

Choosing a plant that is already healthy will start you out on the best foot. Plants don't have to be perfect-looking to be healthy, and even really sad plants can be revived, but for a head start, look for these things when plant shopping:

ALL THE PLANTS SHOULD LOOK HEALTHY. Since many plant problems can spread, if most of the plants look sick or sad, you may be bringing home a plant problem without knowing it, even if you manage to find one good-looking one.

NO BUGS. Get your face close to the soil and squeeze the pot to inspect for tiny gnats (fungus gnats). Inspect the leaves for other pests, like mealy bugs or spider mites.

NO GROSS ODORS. Quality potting soil doesn't smell foul; it may smell "weird," like earth or plant fertilizer, but shouldn't smell rotten or like manure.

JUST-RIGHT SOIL. If soil is too dry, the plant may be stressed. Too wet, and the roots may already be rotting. Potting soil should ideally be damp to partially dry when you purchase your plant, unless you can see it's been recently watered, which would cause all the plants to feel wet.

ROOTS ESTABLISHED. Gently tug on the stems at the base of the plant closest to the soil; if there's resistance, the roots are established; if the plant comes out of the soil, it may not be rooted properly. In that case, look for roots, and if you don't see any or they're very small, move on to the next plant.

PLANT INFORMATION. If the plant doesn't have a tag, see if someone at the shop knows what it's called and how to care for it. If it does have a tag, check to make sure the tag matches the plant by doing a quick internet search. Knowing what species the plant is will be your biggest help in finding out what it needs and caring for it well. Use your phone or ask the people working at the shop (if they seem knowledgeable) to determine the optimal conditions for the plant you want to buy, and make sure it matches your home environment.

OTHER PLANT PROBLEMS: See our Plant Problems guide, pages 164–179.

LIGHT

We've already covered the essentials of understanding light earlier in the book (see pages 4–6), but keep these additional tips in mind as you choose your plants:

- When selecting a plant, think about its light needs first. Watering and humidity needs are much easier to accommodate, but if you don't have enough light, it'll be difficult to work around that, and your plants will suffer. Always be realistic rather than aspirational about a light assessment of your home.

- Know how the light changes in your space. Just because your living room is bright in the morning doesn't mean that room stays bright. A window that gets indirect light most of the day might get blasted with direct sun in the late afternoon. Take time to observe the light at different times of day in any room where you plan to put plants so you can select the right ones.

- Make sure you understand light terminology to interpret care instructions. Here's a review of commonly used terms and their meanings:

DIRECT SUN OR FULL SUN = plant needs to be in the path of the sun (remember, if the plant had eyes, could it see the actual sun?) and should get 4 or more hours of direct sun each day.

BRIGHT INDIRECT LIGHT = plant needs a bright spot, usually within a few feet of a window, where there are 6+ hours of light (but not direct sun) each day.

MEDIUM OR LOW LIGHT = plant can tolerate a spot farther away from a window with no direct sun. (Note: there are very few plants that require this kind of light indoors.)

FULL SHADE = this term often appears on plants that are mainly intended for outdoors. With indoor plants, full shade is equivalent to bright or medium indirect light.

PARTIAL SHADE OR MIXED SHADE = again, usually used for outdoor plants. The indoor equivalent would be a mix of direct sun and bright indirect light.

WHAT TO EXPECT WHEN YOU BRING YOUR PLANT HOME

Most houseplants are grown in commercial greenhouses where bright indirect light is abundant, humidity levels are optimal, and automated systems manage watering and fertilizing plants. Compare that environment to your home, and it should become clear why a plant might go through some transition as it adjusts to life in your home.

Some effects of transition you're likely to see in the first few weeks after bringing a plant home:

- Leaf drop (i.e., leaves falling off a plant).
- Yellowing of leaves, especially smaller leaves that get shaded out by larger leaves.
- Slight drooping of the whole plant, even if the plant was recently watered.

Don't fret—all of these symptoms are normal and to be expected as the plant adjusts to lower light and humidity levels and a new watering and fertilizing routine. After 2–3 weeks with proper care, the plant should be settled in, and you shouldn't continue to see these issues. If they persist for more than a month, assess your care and ensure you're providing everything the plant needs to be happiest.

Keep in mind that a plant you've just purchased doesn't immediately need to be moved to a new pot. The plastic grower pots that plants come in will be a fine home for a plant, often for many months to a year. Recently purchased plants may even do better with more time to establish their roots in the soil they come in. If you do decide to move your new plant to a new pot, use the tips on pages 137–142 to determine the right size pot.

CHOOSING YOUR POT

You've picked out the perfect plant and know just where you want it—you're ready to put it in a pot/planter (these two terms are used interchangeably) right? But which pot is the right one? There are, of course, tons of options, but not all of them will be right for you.

After years of seeing people kill plants, I've come to believe that one of the most important qualities in a planter is a drainage hole, especially if you're new to plant care.

- Without a hole for drainage, there is no way for excess water to escape, so even a small amount of water beyond what the plant can use quickly means the plant's roots may be sitting in standing water, which suffocates the roots, causes rot, and attracts pests.

- The "drainage layer" technique of layering rocks and/or charcoal at the bottom of a pot with no hole may work for some people, but you have to be very careful to only use as much water as the soil can absorb and the plant can use quickly, which is difficult to guesstimate. It's not a method I recommend, especially for beginner plant people.

If your pot doesn't have a drainage hole, there are ways to work around it and still keep your plant healthy.

Cachepot

1. Keep the plant in a grower pot (the pot, usually plastic, that your plant came in) that has drainage holes.

2. Place the plant inside a decorative pot that is slightly larger.

3. Remove the plastic pot from the decorative pot before you water it, then allow the plant to drain at the sink or tub.

4. Return plastic pot to decorative pot.

DESIGN TIP! **IF YOUR DECORATIVE POT IS** too deep to properly display the plant (leaves sit too deep inside the pot), you can use pebbles or a plastic jar lid to raise the grower pot up a bit. If it's *way* too deep, save packaging Styrofoam (don't buy new) to fill the bottom of the pot, placing the plant on top.

Add Drainage Holes

It's easy to add drainage holes to most planters or even non-planter vessels, like mugs or vases, as long as they are made of ceramic, clay, porcelain, or pure concrete.

YOU'LL NEED:

- An electric drill plus a bit called a diamond coated (or diamond grit) hole saw for masonry. These bits come in a range of diameter sizes. A ½" works well for most planter sizes.
- Water (in a small pitcher or watering can with spout is easiest).
- Small towel to catch excess water.
- The planter you're adding the hole to.

1. Spread towel out over your work surface. Place planter on towel with the bottom end facing up. Decide where you want your hole or holes to be. Smaller planters usually only need one hole in the center. Add more holes as you go up in planter size to ensure water can drain quickly. Space holes evenly.

2. Pour some water over the surface of the bottom of the planter (enough to cover it and preferably pool in the middle). This will reduce friction as you drill and minimize any risk of the pot cracking. Begin drilling on a slow speed, keeping the bit at a slight angle as you start to minimize slipping.

3. Once the bit has caught on the surface and starts to make a channel into the planter, straighten the bit out until it's level with the surface and increase your drill speed. Apply even, light pressure—just enough to maintain contact with the pot as the drill works through the planter material.

4. As the channel you're drilling gets deeper, you'll eventually feel a change in resistance as the bit gets closer to breaking through. Keep good control so you're prepared for the moment it does. The time it takes to break through will vary depending on the power of your drill, the material you're drilling, and the sharpness of the bit (older bits that are worn down will take longer).

5. Once the drill breaks through, you'll have a perfect hole. Rinse off any residue on the planter and bit. Dry bit and drill thoroughly after each use.

Types of Pots

Beyond drainage holes, there are many types of pots to choose from, and each has different benefits and downsides.

PROS OF PLASTIC

+ Lighter weight. A lighter pot makes moving it to water easier, and it's better for hanging if you're concerned about the weight.
+ Holds more moisture in. Since plastic isn't porous, less water is lost from the soil through evaporation, which means you'll need to water less often.
+ More drainage holes. Most plastic pots, like the grower pots plants come in, have drainage holes all along the bottom of the pot. More drainage holes can help prevent overwatering, since there are many places for water to escape from.

CONS OF PLASTIC

− Because plastic pots can hold in more moisture, if you accidentally over-water and need a plant to dry out quickly, non-porous plastic may slow down the drying-out process.
− Environmental impact. Plastic pots, if not reused, are becoming difficult to recycle since many are made from plastics that aren't being recycled anymore.

PROS OF CERAMIC

+ More stylish. Ceramic pots are usually more aesthetically pleasing and come in a wide array of colors, textures, shapes, and sizes. No matter your decor taste, you're likely to find a ceramic pot that fits your style.
+ Can hold moisture in if glazed on the inside. Similar to plastic pots, if your ceramic pot is glazed on the inside, it will hold more moisture in, which can reduce watering frequency.
+ May be more environmentally friendly, depending on the source. Ceramic often lasts a long time, but even if these pots end up in the trash, ceramic eventually breaks down, so it's a more eco-friendly choice. Ceramic pots from larger factories may still be made with practices and materials that aren't eco-friendly, though, so do your homework if you're concerned about that.

CONS OF CERAMIC

– Lack of drainage holes. Many ceramic planters don't already have drainage holes, so you have to add a hole or use it as a cachepot.
– Heavier. Heavy pots can be difficult to move when watering and pose a problem for hanging plants if the pot is too heavy for the hardware you're using to hang it.
– More fragile. Ceramic pots can break more easily, which may be of special concern to those with pets or kids who could potentially knock pots over.

PROS OF TERRA-COTTA

+ Can be stylish. Terra-cotta appeals more to some people than others, but generally, these pots can be used in a range of decor styles, and in curated groupings can look very stylish.
+ Porous. Unglazed terra-cotta is porous, so air can flow through and roots get more oxygen. The porousness also makes it easier for plants to bounce back from overwatering, since the soil can dry faster.

CONS OF TERRA-COTTA

− Dries out very quickly. The flip side of terra-cotta's porous nature is that the soil will lose water through evaporation faster, which leads to watering more often.
− Mineral buildup. Water contains mineral salts, which build up easily in the porous clay, leaving a white residue on the outside of the pots that some people dislike.

Choosing the Right Size

The size of a pot is just as important as the kind of pot. A planter too large for the root system of a plant can essentially drown the roots in wet soil, since the plant isn't big enough to use all the moisture the soil can hold.

The best rule of thumb is to go up no more than 2" in diameter at a time whenever you move a plant to a bigger pot, so if your plant is in a 4" pot, your new pot should be no larger than 5–6". The depth of the pot should be a good fit for the plant. Plants like cacti or succulents, which have finer, shallow roots that stay near the surface of the soil to access even a small amount of water seeping in, will do well in a shallow bowl-style planter. Most other plants will need something deeper, like a more classic shape of planter, so their roots have enough space to spread out and grow.

CHOOSING YOUR SOIL

So you've picked a pot or planter. If you'll be keeping your plant in a grower pot and using a cachepot setup, you can simply leave the plant in the pot it came in until the plant has grown and it's time to re-pot it. If you're not using a cachepot, consider keeping the plant in the grower pot for a month or so, so it can adjust to life in your home. Once the adjustment is over, you're ready to pot.

Just like different plants have different light and water needs, they also have different preferences for what medium (e.g., soil or other material) they are planted in. Some of the common mediums you'll encounter are:

ALL-PURPOSE MIX: This mix, typically made of some combination of peat moss, bark, compost, perlite, pumice, and fertilizer, will work for many indoor plants.

CACTUS/SUCCULENT POTTING MIX: Best for plants like cacti, succulents, ZZ plants, and *Sansevieria* species that prefer to dry out quickly and need faster-draining soil. This mix usually has less or no peat moss and a higher percentage of perlite or pumice, in addition to some sand.

SPHAGNUM MOSS: Some plants, called epiphytes, don't need soil at all. In the wild they attach themselves to trees where leaves and other material from the trees above will keep the roots moist, but if you're keeping these plants indoors (where the air is drier), planting (or mounting them) with sphagnum moss will help keep the roots moist.

As you become a more experienced plant person or find you're focusing on a certain genus of plant, you may want to develop your own customized potting medium by amending a soil mix (i.e., adding other materials to it) to create a more customized mix. Some of the components you can add to soil are:

ORCHID BARK OR CHUNKY COCO COIR CHIPS*: Helps with drainage while still retaining some moisture, allowing more oxygen to reach the roots.

PERLITE OR PUMICE: Helps mix to drain faster, stores water and releases it slowly, makes soil lighter (useful if you have a lot of pots to carry around), prevents soil from becoming compacted.

PEAT MOSS OR FINE COCO COIR FIBERS*: Helps mix retain more moisture while still providing drainage and allowing more oxygen to reach the roots, prevents soil from becoming compacted.

ACTIVATED CHARCOAL (A.K.A HORTI-CULTURAL CHARCOAL): Absorbs toxins, helps kill bacteria/fungi, can help prevent root rot.

SAND: Usually added only for cacti/desert plants to improve drainage.

*Coco coir is a by-product of coconut fibers, harvested during coconut processing. Unlike peat moss, it is a renewable resource and considered more sustainable.

Healthy Soil Is Important for Healthy Plants

How do you know if your soil is unhealthy? Here are a few clues:

SOIL IS SUSPICIOUSLY CHEAP. Cheap soil usually has low-quality ingredients, which will not be helpful to a plant.

SOIL IS EXTREMELY HEAVY FOR THE SIZE OF THE BAG. Heavy potting soil is usually either very wet, which can cause fungus or attract gnats, or has high sand content, which is not helpful for most plants.

TINY GNATS (CALLED FUNGUS GNATS) FLY OUT OF THE BAG. This is common even in a brand-new bag of potting soil because some common ingredients provide good hiding places for these gnats, and soil often sits at garden centers for a long time, giving gnats time to find their way in. Soil with visible adult fungus gnats will contain even greater numbers of larvae, which will be trouble if you use that soil to pot your plant, since all those larvae will hatch out in your home! When you open a new bag of soil, do it slowly, and keep your face close to the opening to look for gnats trying to fly out. If you discover your soil has gnats, you'll either want to return it or use it for outdoor plants instead (outdoors, predatory insects or birds will eat the gnats). If you've already used soil that has them, see page 175 for tips on dealing with fungus gnats.

Personal Potting Mix Recipes

DANAE HORST
Folia Collective

FAVORITE TYPE OF PLANT: *Hoyas.*

POTTING MIX RECIPE: 2 parts peat or coco coir-based potting mix, 1 part orchid bark, ½ part horticultural charcoal.

WHY I PREFER THIS MIX: The peat/coco coir helps retain some moisture, while the orchid bark adds more air pockets to the mix, helping roots get more oxygen. The addition of charcoal prevents bacteria to keep roots their healthiest.

Photo: Larry Dieterich

ENID OFFOLTER
Aroid grower/owner of NSE Tropicals

FAVORITE TYPE OF PLANT: *Philodendrons.*

POTTING MIX RECIPE: ¼ Pro-Mix potting mix, ¼ small charcoal chunks, ¼ Orchiata orchid bark, and about ⅛ perlite and ⅛ tree fern substrate.

WHY I PREFER THIS MIX: Nice crunchy mix that drains well to help keep the roots healthy and encourage growth.

Photo: Becca Stevens

JEN TAO

Succulent enthusiast known as @jenssuccs online

FAVORITE TYPE OF PLANT: Succulents!

POTTING MIX RECIPE: To be totally honest, I will often buy premixed cactus soil. There are some very high-quality ones out there, and for me the ease of purchasing one bag beats out buying many different components and making my own. If you have a hard time finding a good mix or just enjoy the process of getting your hands dirty, then I suggest mixing one part potting mix to one part pumice. Try to use a good organic potting mix that doesn't contain added fertilizer; in my opinion, succulents really don't need it.

WHY I PREFER THIS MIX: Ease and effectiveness. A good-quality potting mix and pumice are easily found at your local garden store. The pumice makes the soil less dense and faster drying, which is key for healthy succulents. You can find much more complicated soil recipes online, but I believe that the more important issues are just making sure that you have them planted in the proper pot—drainage is key! And be sure you are watering correctly: water thoroughly, allow soil to dry completely, water thoroughly, repeat. If you take care of those two steps, your plants should be very happy even in a simple succulent soil!

Photo: Aaron Keeny

HILTON CARTER

Author of Wild At Home

FAVORITE TYPE OF PLANT: I prefer a good *Ficus*! A fiddle leaf fig, rubber tree, or council tree is always welcome in my home.

POTTING MIX RECIPE: I don't get too fussy with my soil for my tropicals. I like to take an organic all-purpose soil and mix in some perlite. The mixture is 7 parts soil and 1 part perlite. I just recently started to add horticultural charcoal to the bottom of all my pots. I fill the bottom ⅛ of the pot with charcoal and then add my soil and plant.

WHY I PREFER THIS MIX: I love this mixture for my tropicals because while I don't want them to get dry too fast, my goal is to always make sure their soil doesn't stay damp. The addition of the perlite helps to absorb some of the extra water.

POTTING

Whether it's a new plant you've just brought home or one you've had for a while, a houseplant will eventually run out of room to grow in the pot it starts out in.

Repotting, Potting Up, and Potting On

These terms are often used interchangeably and may mean different things depending on where you live, but they do indicate a few different potting methods.

REPOTTING: Cleaning up a plant and changing out its potting mix without moving it to a larger pot. This is a method that refreshes the potting mix to add fresh nutrients for the plant and clear out any soil that's become laden with mineral salts.

POTTING UP: This usually refers to giving a seedling (plant grown from seed) or a rooted cutting a pot for the first time.

POTTING ON: Probably the most common method, where a plant is moved from one pot to a larger pot.

These methods may be used at different points in a plant's life, but the essential techniques involved are still the same.

Proper Potting

People often confess to me their fear of potting, and while I try to reassure them, I understand that it still seems like a daunting task. It really isn't something you need to fear. Potting a plant the right way will help prevent damage to roots, ensure healthy growth, and make the stress of the process easier for the plant to handle—and it's actually fairly easy to do it the right way, when armed with the right info and a few simple tools.

Before you begin the potting process, make sure your plant isn't too wet or too dry—aim for halfway between waterings to make it easy to handle the root ball without risking damage to the roots. You'll also want to consider the proper pot size (see page 129) to determine if you'll be repotting, potting up, or potting on.

Before you pot up a plant, make sure it's the right time to do so. How do you know if it's time to move your plant to a larger pot? Plants often give us many signs that they are ready:

- Visible roots coming out through the drainage holes in the bottom of the pot.

- Visible roots coming out at the top of the pot. (These are not to be confused with aerial roots, which grow above the soil and which plants use for climbing, support, and accessing more moisture and nutrients. Aerial roots are typically thicker than soil roots and will start from the stem or trunk of the plant rather than under the soil.)

- Visible roots pushing against the side of the pot. This is easiest to spot in plastic pots, where the roots pushing against the side will cause visible bulging.

- You're finding you need to water more frequently than normal, even if it's not a warm time of year. As roots overtake the soil in the pot, water runs right past them, and there isn't enough soil to help retain moisture, so the plant needs water more often.

- Lack of growth during a period when you would usually see growth. As a plant runs out of room for its roots to expand, growth will usually slow down.

- It's been more than a year or two since the last potting. Most plants should be repotted or potted up (see below) every 12–18 months.

When you're done potting, if you have potting mix left over, seal the bag up with packing tape and place it in an airtight bin. Store the bin in a dry, cool place. Most potting mix will be fine to use for about six months from when you purchased it. If it's been longer than six months, it may still be fine, just check it well for pests and mold and don't use it if you find either. Plants potted in mix older than six months should be fertilized right away, as long as it's during the growing season. Fertilizer content in most potting mixes degrades over time, usually within the first 3–4 months.

If you've potted an established plant, unless the plant was previously dry or the potting medium you've used is *very* dry, wait a few days to water to allow the roots to spread out as they seek water. This will help the plant to more quickly establish its root system in the new soil. Cuttings or seedlings should be watered right away, as their roots are more tender and will dry out more quickly.

Potting can cause stress for a plant. To provide optimal care for your newly potted plant in the weeks afterward, provide appropriate watering, keep the plant in the same light it was previously in, maintain appropriate humidity levels for the plant's needs, and protect it from harsh changes in temperature. Don't panic if your plant loses a few leaves or seems less happy than usual—if you potted it properly and you're providing good care, this is normal as the plant adapts, and after a few weeks, your plant will be happier than ever in its new pot!

WATERING

Water is second only to light in meeting a plant's essential needs. Water also has the power to make or break a plant's health; incorrect watering often leads to the untimely demise of your plants.

MISCONCEPTION: *Plants need to be watered every day.*

TRUTH: Very few houseplants need water every day, and many will actually struggle if they are watered every day. (The exception here is plants potted solely in sphagnum moss, which may need water daily, especially in warm months.)

MISCONCEPTION: *A watering schedule is the best way to determine how often a plant should be watered.*

TRUTH: Though a schedule can be helpful, especially if you're prone to forget about your plants, there are many factors that affect how often a plant needs water: the type of plant, how much light it gets, time of year, type of pot, type of soil, thoroughness of each watering, and whether the pot has drainage. Getting into a regular rhythm of checking on your plants is great, and a schedule is a good place to start, but for the happiest plants, you should check how moist or dry the soil is and water only when the plant needs it. Over time, as you get to know your plants, you'll start to notice all the little signals they're giving you

when they need water, even without a scheduled check-in (see When to Water on page 146 for more details).

MISCONCEPTION: *Letting a plant "dry out" means drying out on top only.*

TRUTH: How deep you'll want to check for dry soil will vary from plant to plant. Some plants need water when just the surface of the soil begins to dry out, others should dry out 1–2 inches, and others still should dry out all the way to the bottom of the pot.

MISCONCEPTION: *Plants that like to be dry should only be given small amounts of water at once.*

TRUTH: Soil should be fully saturated every time you water, no matter the plant. What will vary from plant to plant is the *frequency* of watering. Deep waterings ensure the whole root system has equal access to water. When only small amounts of water are given, water tends to saturate only parts of the soil, leaving some portions of the roots in need of water and others oversaturated.

When to Water

Good care instructions will give you an idea of when to check or when to water, but plants themselves also give you signs they need water:

SOFT OR PALE LEAVES. When leaves don't look as perky, shiny, or vibrant as usual, check the soil and you'll likely find it dry. Once watered, your plant will be back to normal within a day.

STEMS FLOPPING OVER. If you ever find a plant completely wilted, as long as some of the leaves and stems are still green, give it a thorough watering and it will usually perk right back up after a day.

SOIL IS PULLING AWAY FROM THE EDGE OF THE POT. Soil that has gotten too dry will shrink as the water dries up, and if it's allowed to dry too much, that shrinking becomes visible. Once a plant is at this level of dryness, it will need a serious soak in order to bounce back (see bottom watering method on page 151).

Types of Water

I get a lot of questions about what type of water plants prefer. While I don't know exactly how a plant would answer that question, my guess would be rainwater, since that's how plants are watered in the wild. However, since most of us can't consistently provide rainwater for our plants, we can make use of the options available to us:

- Most plants will do fine with **tap water**, as long as you don't have a water softener (which produces sodium-laden water) or excessively hard (mineral-heavy) water. If you do have hard water, installing a reverse osmosis or charcoal water filter system can help make your water easier on your plants.

- **Room-temperature water** is best—cold water can shock plants; hot water can damage sensitive leaves or roots.

- Some plants, like *Maranta*, *Calathea*, and *Tillandsia* species, as well as carnivorous plants, do best with **distilled water or rainwater**, which have fewer mineral salts than other water. All plants will be happy with either of these types of water, but for most, the extra effort isn't necessary, unless you have easy access to one or both.

Watering Methods

There are a variety of watering methods that work well, but you'll want to keep these basics in mind, no matter the method:

WATER SHOULD ALWAYS BE APPLIED EVENLY across the entire surface of the soil. This ensures all of the roots get equal access to water and you don't end up with dry pockets or soggy pockets of soil. Move the water source (or move the planter) as you water in order to achieve this.

WATER UNTIL THE SOIL IS FULLY SATURATED. This will often mean that water is flowing out of the drainage holes, and if it's a planter you can pick up, it will feel quite a bit heavier.

IF YOU'RE NOT SURE IF YOU'VE GIVEN THE PLANT ENOUGH WATER, use a moisture meter and check in a few different spots. Moisture meters range from simple to high-tech, and most can be found at any garden center or plant shop. I find the simplest type works just fine, giving a quick reading of how wet or dry the soil is when inserted into the soil.

The watering method you choose depends on a few factors: how you have your plants arranged, the type of pots, the type of soil, and personal preference. The most common watering methods are:

WATERING CAN

- Best for: large plants that can't be moved, planters with deep saucers that can hold excess water as it drains out, and hanging plants that can't be taken down (and have a saucer to catch excess water).

- There are a wide variety of watering can styles out there. I prefer a can with a long, narrow spout that comes out at the bottom edge of the can, which can be used to direct water more precisely and give easier flow control without having to tip it at an extreme angle. No matter the style, pick a can that, when filled with water, is light enough for you to lift, carry around the house, and easily pour from.

- When watering plants in place, pay careful attention to how much water is filling the saucer to avoid overflowing it. Water may run down the side of the pot and directly into the saucer without wetting much of the soil, so you may need to pause on watering, let the plant draw some of the excess water back up into the soil, and then return to finish saturating all of the soil.

- Once you've given plants a few hours to draw any excess water back up into the soil, empty remaining excess water from the saucer. For large plants that can't be moved, use a turkey baster to remove the water or a terry cloth towel to soak up the excess.

WATERING AT THE SINK/TUB

- Best for: plants in cachepots (just be sure to remove the grower pot from the cachepot before watering), large quantities of plants that all need water at once, hanging plants, and plants that need a good "shower" to clean dust off leaves.

- This method is easiest with a sprayer-type faucet. For my houseplants, I purchased a detachable sprayer for our showerhead specifically for watering plants.

- For hanging plants, use your showerhead to water, then let drip-dry on the shower curtain rod or on a hanging clothes rack with a towel underneath. Once the plants have stopped dripping, you can return them to their normal spots.

- If you prefer to add fertilizer to your water, you can still water in the sink or tub—just use a watering can instead of the faucet.

BOTTOM WATERING

- Best for: plants in pots with drainage holes, large quantities of plants that all need water at once, plants that have gotten overly dry, and plants potted in moss or peat/coco coir-heavy mixtures.

- To bottom water, simply fill your sink basin, tub, or even a large, deep baking pan (if you only need to water a few plants) with a few inches of water. Place plants in the water and let soak until the soil is fully saturated—usually a few hours will be enough time. Once the soil is saturated all the way to the surface, let the planter drain briefly before returning to a cachepot, saucer, etc. For fastest draining, use a cooling rack (like you'd use for baking) to elevate the plants and allow the water to drain out without pooling around the bottom of the pots.

- If you usually add fertilizer to your water, simply add enough for the volume of water you're using to bottom water.

MISTING

- Best for: in-between deep watering for ferns, mounted plants, and air plants.

- As I mentioned in the humidity section (pages 86–87), misting is not a very effective way to provide long-term humidity, but it is helpful for certain plants to provide moisture in-between their usual waterings.

- Cute metal misters are fun to display, and well-made versions, like the ones made by the company Haws, produce a perfect superfine mist. Any spray bottle that creates a fine mist will also work well for this purpose.

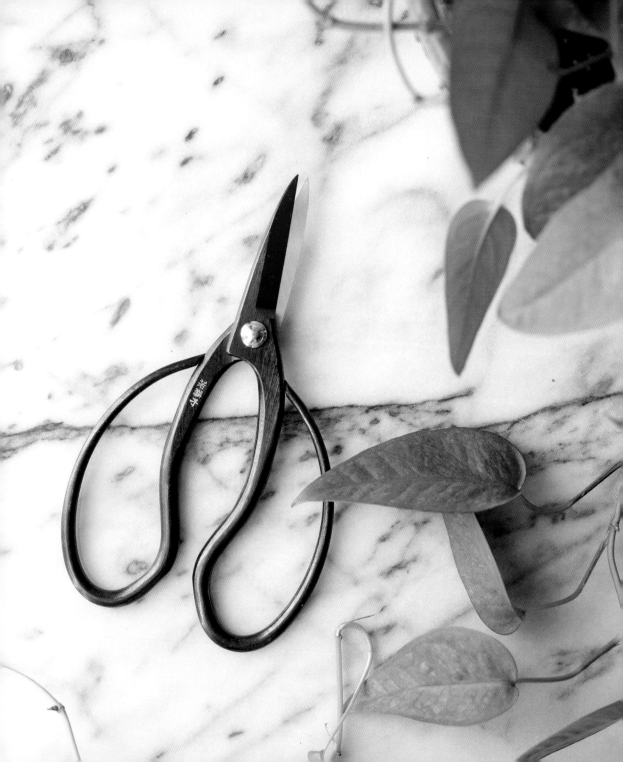

PRUNING

Pruning is another task that can feel very scary but is actually pretty easy and very important for managing the shape of a plant and encouraging new growth. Plus, it yields cuttings for propagation—an added bonus!

There are several reasons to prune a plant:

- The plant has become unruly in size/shape.
- The plant has become stretched out (etiolated) over time with new leaves that are smaller and smaller as the plant reaches for more light, creating a sad-looking plant.
- The plant has had pests or other problems that have produced unattractive/dying leaves.
- You'd like to see branching and/or a fuller plant.
- You'd like to use cuttings to fill in a plant that has gotten thin on top.
- You'd like to share your plant through cuttings.

How you prune a plant will depend on the kind of plant, but some general guidelines to follow are:

- Prune or pinch during active growth (when new leaves and shoots are growing). In most growing zones, this will be in the spring and summer.
- Always use clean, sharp tools, taking care not to crush the stems. Pruning shears or a sharp knife are best. If you use household scissors, ensure they are sharp and sterile (clean with rubbing alcohol or sterilize with heat).
- Cut or pinch above new leaves or nodes (the little bumps that stick out along stems) at a 45-degree angle.
- Pinching (see page 155) is best for producing denser growth.
- Cutting back (see page 154) is best for reshaping a plant or restarting a languishing plant.

Most Common Pruning Methods

CUTTING BACK

- Helps reshape a plant, redirects growth, restarts languishing plants.

- Use pruners, shears, or a knife. Always ensure cutting tools are clean and sharp.

- Cut above new leaf or node at a 45-degree angle.

- Good plants for this method:

  - *Ficus* species or other woody-stemmed plants
  - *Philodendron* species
  - *Monstera* species
  - *Epipremnum* species

PINCHING

- Helps produce denser growth.

- Use thumb and forefinger (or a knife/shears) to pinch back just the growing tip.

- Remove ¼ to ½ inch of the growing tip above a node.

- Pinch back as soon as active growth begins (i.e., when you see new leaves and shoots), and continue through the whole growing season.

- Good plants for this method (though not all have care info included in this book):
  - *Fittonia* species
  - *Tradescantia* species
  - *Nematanthus* species
  - Herbs

FERTILIZING, A.K.A. FEEDING

Beyond the light needed to photosynthesize, plants also need essential nutrients to thrive. In the wild, these nutrients would be supplied through organic matter, like leaves, animal droppings, etc., that would break down into the soil or water. When plants are grown in cultivation, we have to supply these essential nutrients through fertilizer.

There are three chemical elements most essential for plant nutrition: nitrogen, phosphorus, and potassium. Fertilizer packages will indicate the ratio of these elements as a number like 7-10-7, which refers to the relative content of nitrogen (N), phosphorus (P), and potassium (K), in that order.

Most houseplants will do fine with a balanced mixture like 10-10-10, but there are fertilizers made with specific kinds of plants in mind. Generally speaking, fertilizers with more nitrogen are best for leafy plants, so a slightly higher first number is especially good for most houseplants.

Feeding houseplants is most important during the active growing season. If you are seeing new leaves or shoots, it's safe to fertilize that plant. Again, in most growing zones, this will be during the spring and summer. Once growth slows down, you can stop feeding until dormancy is over and new growth has started to come back. How frequently you feed will depend on the particular fertilizer you're using, so always follow the directions on the package.

It *is* possible to overfeed a plant and risk damaging the roots or leaves, and following the directions on the package of fertilizer is essential to avoiding this. When in doubt, go with a more diluted mixture to avoid these issues.

Forms of Fertilizer

LIQUIDS OR POWDERS that are mixed into water and applied during plant watering. These are easy to use and can be diluted to make an extra-light concentration—useful for sensitive plants or if you don't know when the plant was fed last.

GRANULES OR SMALL SOLIDS that are mixed into soil and made to be a slow-release fertilizer over a set period of time. These are very convenient and work well as a soil additive, often appearing in pre-made potting mixes. If you see small perfectly round colored balls in your potting mix, these are likely fertilizer pellets.

STAKES that you place in the soil. These are also slow-release and dissolve over time. These are very convenient but have a risk of over-concentrating fertilizer in one area of the soil, potentially damaging the roots there.

FOAMING LIQUIDS that can be applied directly to the soil or mixed into water. The direct-to-soil application has the same risks as the stakes, as it's harder to evenly apply.

FOLIAR SPRAYS that can be applied directly to the leaves. This type of fertilizing is good for plants that need an immediate boost, like plants that have been starved for nutrients for a long period. This is also a useful method for epiphytes, like air plants and many bromeliads.

Which type of fertilizer you choose will depend on what kind of plants you're feeding, how you water your plants, and other lifestyle factors, like your time and desire to track your fertilizing. Read the packaging when you're picking out fertilizer. The labels will indicate what kind of plant the product is for, how to mix it (if needed), and how it should be applied. Follow the directions carefully to avoid fertilizer-related problems, like damage to leaves or soil.

CLEANING

In the wild, rain and wind help to clean plant leaves, but in our homes, they need a little help. If too much dust builds up on the leaves, it can hinder the photosynthesis process and make it harder for the plant to "breathe" through transpiration. To help your plants with this, make dusting/cleaning leaves part of your monthly plant routine.

- For most leaves, use a soft, slightly damp cloth. Gently wipe away dust with the cloth while supporting the leaf with the palm of your other hand to avoid ripping or otherwise damaging the leaf.

- If hard water builds up on leaves, try wiping them with diluted white vinegar to gently break down the mineral deposit. Some people also swear by milk or mayonnaise for this purpose, but I find diluted white vinegar gets the job done with less risk of attracting pests.

- A note on "shining" leaves. While there are many products out there that make plant leaves look shiny, most contain heavy oils, wax, or silicone, and those can have negative effects on the leaves. If you like a shiny leaf, try using neem oil when you clean your leaves. A natural oil made from the neem tree, neem oil is used to treat many common plant problems, repels pests, and will leave a subtle shine but is easier on the leaves than products sold as leaf shine. You can find neem oil at most garden centers sold in various forms, some of which require mixing with water, some ready to use. Follow the directions if the label indicates the neem needs to be mixed with water.

PROPAGATION

As your plant knowledge grows, your plant collection likely will too. One of the easiest ways to add to your collection is propagation, or growing a new plant from part of a "parent" plant. Propagation is also a great way to share plants you love with friends and family.

There are many ways to propagate, and the type of plant you're working with will dictate which method you use, but for beginners, water propagation is easiest. This method works for a variety of plants and allows you to see new roots as they develop so you know if it's working or not.

- Use a clean, sharp knife or shears, taking care not to crush the stem as you cut.

- When taking cuttings of most leafy plants, find a portion of your plant that is flexible enough to still bend but not so tender it snaps when bent—essentially somewhere between the oldest growth (which can be woody or very tough) and the newest growth (which is often too tender). This will give you the best chance of seeing roots form. Now find the "nodes," which are the little bumps that stick out along the vine or stem. Using your shears or knife, cut cleanly just below the node or branching point. This is the spot where the new roots will form. To ensure success, make sure you have at least two nodes on the cut portion of the stem—this way if one fails, you can cut again and try once more.

- Start with a clean vessel. Amber glass (or other dark-colored glass) is helpful since the water will stay algae-free longer with protection from light. If you're using clear glass, you'll just need to change the water more frequently.

- Remove any leaves that will be below the water level once the vessel is filled. Leaves will decompose in the water and affect the health of the cutting.

- Fill your vessel with enough clean water to reach the lowest nodes. Place in an area that gets indirect light but no direct sun. Monitor the water level to make sure the nodes are kept in water. Change water once a week or so to keep bacteria at bay and ensure healthy roots.

- Roots should form in anywhere from two weeks to a few months. Be patient, but check occasionally to ensure the stem isn't rotting. If it does rot, trim that portion away, change the water, and try again. Once roots form, if you plan to transplant to potting soil, wait until they're about an inch long, then move to a small pot of lightly moist soil. If you'd prefer to keep your cutting in water long-term, known as "hydroculture," let the roots keep going as long as the vessel can contain them. Just remember that you'll need to add liquid plant food to the water occasionally to add nutrients that would normally come from the soil.

This propagation method works for a variety of plants, including:

- *Philodendron hederaceum*
- *Epipremnum aureum*
- *Scindapsus pictus*
- *Hoya* species
- Many *Peperomia* species
- *Monstera deliciosa*

To learn more about other propagation methods, see the recommendations in Resources (pages 186–187).

PLANT ON!

Empowered with the essentials of plant care, you'll find many of your plants look happier and healthier than when you were just winging it! Referring back to these essentials in the future can often help you troubleshoot issues and adjust your care to help get your plants back on track. Running into problems these essentials don't address? The next chapter is for you: Let's tackle plant problems.

Plant Problems

IF A LIFE OF LIVING WITH PLANTS plus years of answering thousands of plant-related questions has taught me anything, it's this: no matter if you have one plant or many, are just bringing your first plant home, or have cared for plants for years, problems will arise. Even plants with the most attentive caregivers can develop pest problems, contract diseases, or suffer periods of neglect when other priorities take precedence. The good news is that most plant problems are fairly simple to solve—especially if you catch them early.

You certainly don't need to go around looking for a plant problem under every leaf, but taking the time to familiarize yourself with the common problems will equip you to quickly recognize them if you see them. When you do find a problem, don't freak out; instead, use it as an opportunity to learn. Make mental notes (or write it down if that's more your style) about what caused the issue so you are less likely to repeat it.

Watering time is the perfect opportunity to check in with your plants. Look out for: pests; odd leaf colors or textures that might indicate pests or disease (see pages 173–179); signs it's time for a bigger pot (see pages 138–139); and overall health indicators, like soggy or compacted soil, yellowing or spotted leaves, or squishy stems.

I've covered a range of the most common plant problems in this glossary, but if you find an issue you don't see here or can't quite identify, take photos and reach out to knowledgeable plant shops or nurseries near you. If you're part of a plant community, whether real-life or online, you'll often find many other people have dealt with the same issues, successfully battled them, and will be happy to share their advice with you.

As you try to identify a plant problem, remember that some symptoms are indicative of *several* issues (indicated with an * in this glossary), so you will need to play plant detective and rule out each by observing your plant and factoring in the care you've been giving it in order to determine which problems you're dealing with.

Underwatering

SIGNS: Soft, wilted, or curled leaves; drooping stems or whole plant slumped over; soil visibly dry, hard, and/or shrinking away from the edge of the pot.

TREATMENT: A severely underwatered plant will need to be bottom watered, usually for several hours, until the soil feels damp and supple again. Humidity-loving plants will perk back up faster if they are also kept in high humidity while the soil is rehydrating. It may take anywhere from a few hours to a few days for a plant to look back to normal after an underwatering incident. Some leaves may have died off and won't ever come back to life—simply trim them away.

PREVENTION: Utilize a consistent watering routine, checking on your plants at regular intervals to ensure you're watering as soon as they need it. If you consistently see watering-related problems in your plants, get a moisture meter to help take the guesswork out of watering.

Excess Water

SIGNS: Yellowing leaves*; water pooled at bottom of cachepot or saucer even if you didn't recently water; visibly soggy soil with water squishing out when you press on it; leaf drop (leaves falling off); leaves quickly turning brown or black in spots*; stems that feel squishy or look black*; foul odor coming from the root ball; visibly black roots*.

TREATMENT: Depending on the extent of the damage, sometimes just emptying any excess water from the pot/saucer and letting the plant dry out before you water it again will be enough. If the soil is very waterlogged, it's best to remove the root ball from the pot and wrap it with a dry towel to wick away as much moisture as you can. Allow the root ball to sit out of the pot a few days to continue drying out. If the roots are turning black or mushy, remove as much soil as you can and trim away the affected roots. Repot in fresh potting mix. A plant won't be salvageable if it has been sitting for too long in more water than the roots can use. Plants this far gone will usually look totally collapsed or all the stems will be mushy. If any healthy tissue remains on a plant like this, you can always take cuttings of the healthy part and try to propagate it as a means of saving it.

PREVENTION: Often this problem is referred to as "overwatering," which can be a misnomer, since it's not just about how often or how much you're watering but rather about the balance of how much water the plant can use compared to how much water is being stored in the soil. To keep those in balance, ensure you are watering only when the plant needs it by checking the soil before

Underwatering

Excess Water

you water and following the relevant care guide for that plant. Keep the plant in a pot with drainage, or make sure the plant has had enough time to drain before returning to its outer decorative pot. Make sure you're accounting for light levels in your watering routine. For example, plants kept in light that's too low can't use water as quickly as the same plant kept in bright light, so they are often "overwatered." As I mentioned above, if you consistently see watering-related problems in your plants, a moisture meter can help take the guesswork out of watering.

Uneven Application of Water

SIGNS: Soil stays wet in patches, dry in others; portions of the plant die from the roots up while other portions are perfectly happy*; parts of the plant wilt and need water before the rest.

TREATMENT: Bottom water the plant to get any dry patches well-hydrated. If any portions of the plant have died off, remove

them, along with any rotten roots. If a significant portion of the plant has died, you may need to transfer the remaining plant to a smaller pot, since it won't be using as much water. Use the 2" rule of thumb—the pot should be no more than 2" bigger than whatever is left of the root ball.

PREVENTION: Apply water evenly across the surface of the soil when watering. Rotate the plant as you apply the water to ensure water is reaching all areas of the soil. Compacted soil (when soil has dried so much it's pulling away from the side of the pot) can cause water to just run down the sides of the pot and not reach most of the root ball. If you see this, bottom water until soil is rehydrated and can properly absorb water again.

LIGHT-RELATED PROBLEMS

Lack of Light

SIGNS: Yellow*, pale*, and/or very small leaves; stems become very thin and stretched out; new leaves are spread out much farther along the stem; plants with patterns look less defined; variegated plants begin to revert back to green.

TREATMENT: Move the plant to brighter light (do this gradually so as to not shock it). If parts of the plant have gotten so leggy (thin and stretched out) that the plant no longer looks attractive, pruning off the leggy portions will help manage the shape and encourage new growth and branching.

PREVENTION: Keep plants in the appropriate amount of light for their needs; remember that no plant will truly thrive in low light, and there are only a few who will tolerate it.

Lack of light

Overly intense light

Overly Intense Light

SIGNS: Brown or bleached-out leaves, sometimes with crispy patches*; strange spotting from sunburn; plants dry out quickly and look wilted constantly; some plants develop squishy spots, almost like they've been cooked.

TREATMENT: Leaves that have been scorched by the sun will not return to their former glory, but just because a leaf has some sunburn spots doesn't mean it's dead. As long as leaves still have green portions, they can photosynthesize and generate energy for the plant. If the appearance bothers you, simply trim off the burned portions or the whole leaf. Keep in mind that if you trim off a portion of the leaf, browning at the cut edge will develop. You can either keep trimming, eventually cutting away the whole leaf, or just accept the brown edge.

PREVENTION: Keep plants in the appropriate amount of light for their needs. Keep leaves and stems away from the window glass, especially in warmer months, as glass can heat up in direct sun and cause burns. If you're getting too much direct sun for the plants you have, filter the sun with a semi-sheer curtain or swap out your plants for others that prefer direct sun.

Uneven Growth

SIGNS: Plant bends or stretches toward the brightest light in the room; new growth comes in smaller than older leaves; leaves all face one direction.

TREATMENT: If the plant is in an area with lower light, you may need to move it closer to the window or light source. Rotating the plant will help even out the growth.

PREVENTION: Unless you're blessed with skylights, it can be hard to avoid plants adapting their growth to get more light from the closest window. Plants closer to a window that gets bright indirect light most of the day will usually grow more evenly. Rotating plants a ¼ turn once a week can help keep growth more even but won't totally solve the problem for plants that are farther from the window. I say go with it, as long as the plant is getting enough light. In fact, sometimes a plant that has bent towards the light will look even prettier in a space, arching gracefully in just the right place.

Not Fertilizing (Feeding)

SIGNS: Small and/or pale leaves*; new growth seems stunted compared to established plant; lack of growth overall, even in growing seasons.

TREATMENT: As long as the plant is in a season of growth (usually spring and summer) and is otherwise healthy, apply your choice of fertilizer, using the directions on the package. Foliar fertilizers, which are meant to be applied directly to the leaves, will often have a faster effect on the plant.

PREVENTION: Add fertilizing to your care routine, fertilizing during growth periods. Refreshing the potting mixture 1–2 times a year can also recharge the plant with fresh nutrients.

Not fertilizing

Too Much Fertilizer, a.k.a. Fertilizer "Burn"

SIGNS: Visible "crust" of built-up fertilizer salts on the top of the soil (this can sometimes also be mineral salts from water, so if a visible crust is your only symptom, it's probably not fertilizer burn). The other symptoms of fertilizer burn can also be indicative of other plant problems, so try ruling those out first, unless you know for sure you've over-fertilized your plant. Other symptoms include yellowing and/or browning on leaves, brown or blackened roots, leaf drop, and slow growth.

Too much fertilizer

TREATMENT: Remove any affected leaves. If the fertilizer "crust" is present, gently scrape off the top ¼ inch of soil. Flush the soil by letting water run through for several minutes, allow to drain, then repeat 3–4 times to flush excess fertilizer out. Refrain from fertilizing for 1–2 months afterward.

PREVENTION: Follow instructions on fertilizer packaging carefully. If you aren't using fertilizer specifically meant for houseplants, dilute it ¼–½ times further than instructed. Only fertilize during seasons of active growth.

DISEASES

For all plant disease problems, separate affected plants from healthy plants to keep diseases from spreading.

Leaf Spot

SIGNS: Spots caused by fungal infections are brown with a yellow "halo," usually accompanied by tiny black specks. Spots caused by bacterial infections have a water-soaked appearance (wet, dark, almost translucent), and sometimes the yellow "halo" as well.

TREATMENT: Remove all affected leaves, sterilizing cutting tools well afterwards. Apply a systemic fungicide, following the directions on the package.

PREVENTION: Leaf spot diseases are often caused by lack of air flow and by water on leaves (usually from misting). Ensure good air flow and avoid misting to decrease the risk.

Leaf spot

Powdery Mildew

SIGNS: White powdery dusting on leaf surfaces. Usually starts in a small spot, then spreads across the whole leaf.

TREATMENT: Remove affected leaves, sterilizing cutting tools well afterwards. Apply a fungicide. Increase air flow and decrease humidity levels until the plant has been symptom-free for a month or more.

PREVENTION: Mildews are often due to overly moist soil and/or lack of airflow. Keep soil moist but not soggy, and ensure good air flow, especially for humidity-loving plants.

Powdery mildew

Viruses

SIGNS: Strange, distorted growth; yellow patches or streaks on leaves; streaked petals on flowering plants.

TREATMENT: Unfortunately, there are no treatments for viral infections in houseplants. Discard affected plant (do not compost it, as the virus can spread).

PREVENTION: It can be difficult to prevent viral infections, but fortunately it's a fairly uncommon problem. Purchase plants from reputable growers and protect your plants from sap-sucking pests like aphids, as they can transmit viruses.

Viruses

COMMON PESTS

For all plant pest problems, separate affected plants from healthy plants to keep pests from spreading.

Fungus Gnats

SIGNS: Tiny flying insects; often occur in large quantities, flying near damp soil; they have an annoying habit of flying close to people's faces.

TREATMENT: Because fungus gnats can reproduce very quickly, a two-pronged approach is best. Catch adults using sticky traps, found at plant shops or garden centers, placed in the plants you've observed the most gnats around. A number of products can control the larvae in the soil; they are either mixed into soil or applied to the top of it. A common natural control method is to use 1 part hydrogen peroxide to 4 parts water and apply the mixture to the soil until it starts to come out of the drainage holes. Covering the surface of the potting mix with sand can also keep larvae from escaping after hatching. Once larvae are controlled, the life cycle will end.

PREVENTION: As I mentioned in the soil section (pages 130–135), potting soil is often a source of fungus gnats, so check your soil well to avoid infested batches. Avoid allowing soil to get overly moist, and empty all cachepots or saucers of standing water as quickly as possible to remove breeding environments.

Mealy Bugs

SIGNS: Small white fuzzy crawling insects that leave fluffy white substances on stems and other plant parts. Eggs are laid in this excretion, which has a cotton-like appearance. Leaves may have a sticky residue on them, referred to as honeydew, which can attract other insects.

TREATMENT: Treat as soon as you notice even a single mealy bug, as they can spread quickly and be frustrating to eliminate if left untreated too long. Remove all visible adults with alcohol-soaked cotton swabs. Spray leaves and stems with neem oil (see page 159) to control smaller outbreaks. For larger infestations, a systemic insecticide mixed into the soil, which the plant takes up through its root system, is often the easiest way to end the problem.

PREVENTION: Frequent inspection for pests will help prevent infestations from developing too far. Systemic insecticide can be preventatively mixed into soil, but there is no real way to prevent mealy bugs 100 percent.

Mealy Bugs

Spider Mites

SIGNS: Teeny tiny spiders that leave fine webbing on stems and leaves. Because they are so small, you may not notice them until they've begun to affect the plant. Leaves may look curled, deformed, or dull. When looking for spider mites, pay attention to the backs of the leaves, as spider mites often hang out there, near the veins. Webbing is usually found near where the leaves join the stem.

Spider Mites

TREATMENT: Start by removing as many mites as possible with a strong stream of water from a hose or shower spray. Wipe down leaves (front and back) and stems with neem oil, or treat with fungicidal soap or miticide. Increasing humidity levels will also aid in recovery.

PREVENTION: Some plants, like the *Calathea* species, are more susceptible to spider mites than others, so inspecting those plants well before bringing them home is the best prevention. Maintaining proper humidity for humidity-loving plants will protect them from spider mites.

Thrips

SIGNS: Very small, narrow, oblong insect. Juveniles are white in color; adults are tan or dark brown/black. Because of their small size, you'll probably notice their effect on leaves first. Leaves may have silvery streaks or speckles that turn brown over time. Growth may appear deformed.

TREATMENT: Spray with insecticidal soap, following instructions on the label. Repeat this treatment as directed, usually every 4–7 days for several weeks. Neem oil may also be effective as a treatment. Keep affected plants moist to limit damage by these sap-sucking pests.

PREVENTION: Thrips are frustratingly mysterious in their origin, so prevention can be difficult. Well-cared-for plants will always be more pest-resistant, as will plants kept at proper moisture levels and given added humidity.

Scale

SIGNS: By the time they are visible, scale insects are small, domed, and usually white or dark brown in color. They can be pried off with a fingernail or knife and are usually hollow underneath (if you see other bumps that cannot be scraped off, they're probably just part of the plant). Scale secrete honeydew as well, so you may see a sticky substance on leaves or on the floor or furniture directly near the plant.

TREATMENT: If caught early, remove all visible scale insects with a cotton swab or toothbrush soaked in alcohol. If the infestation is more advanced, cut away affected leaves and stems. Insecticidal soap or neem oil applied to leaves and stems may be enough to control the problem. Systemic insecticide mixed into soil is usually most effective long-term.

Scale

Aphids

PREVENTION: Like thrips, scale seem to come out of nowhere, so prevention other than good care is difficult. Checking plants often and thoroughly is the best way to catch problems early before they become full infestations.

Aphids

SIGNS: Small, sometimes winged insects; can be a number of colors: green, yellow, orange, red, beige, pink, and black. Usually found on the underside of leaves or on new shoots, aphids also secrete honeydew, so you may see a sticky substance on the leaves before you notice them on the plant. Leaves may be curled or deformed.

TREATMENT: If caught early, aphids may be cleared away simply by spraying with a strong stream of water from a hose or shower head. If any insects remain, cut away affected growing tips, leaves, and stems. If the infestation is more advanced, use insecticidal soap or neem oil, applied to leaves (front and back) and stems.

PREVENTION: Aphids are common on outdoor plants and can easily hitch a ride indoors on clothing or through open windows. If you notice aphids on your outdoor plants, treat those plants quickly and use caution around them in order to avoid accidentally bringing them indoors. Keep windows closed while outdoor plants are infested.

AFTER READING ABOUT ALL THESE PROBLEMS, it might be easy to feel overwhelmed, but remember that good plant care will prevent many of them, and knowing what to look out for now will only help you catch issues more quickly, which will give you the best chance to win these battles and solve your plant problems.

Pets & Plants

MANY PLANT PEOPLE ARE PET PEOPLE too, and understandably so—they both add a lot to our lives! Living with pets and plants can pose some challenges, but it is possible for everyone to live in harmony with a little extra thought and planning.

A huge number of houseplants are technically toxic when eaten, but won't necessarily cause your pet problems with just one bite. While there are some plants, like Cycads, that are very toxic and best kept out of reach, most pets will be fine, even if they indulge in an occasional houseplant nibble.

PLAYING KEEP-AWAY

There are a few tricks you can try to keep pets inclined to eat plants away from them:

- Give them something they *are* allowed to eat, like cat grass.

- Save them some window space—if plants are taking up all the window real estate, pets are more likely to accidentally damage plants when trying to enjoy their favorite pastime—looking out the window.

- Place plants out of their reach.

  - Plant "shelfies" aren't just popular because they look so cute; they're also a simple way to keep plants out of the reach of your furry friends.
  - Hanging plants are one of the easiest ways to add plants to a room, safely away from curious kitties or playful pups.

On the next page, you'll find a list of the houseplants, all covered in earlier chapters, that are known to be non-toxic for pets. If you're still concerned about which plants are safest, talk to your vet or local ASPCA chapter before bringing a new plant into your home.

COMMON PET-SAFE PLANTS

(and their preferred environments)

- *Peperomia* species (bright indirect/medium), page 48
- *Calathea* species (humid), page 77
- Spider plants (bright indirect), page 46
- *Maranta* species (humid), page 84
- Ponytail palm (direct sun), page 31
- *Hoya* species (bright indirect), page 45
- African violets (humid), page 85
- Palms (bright indirect), page 80
- Ferns (humid), page 81
- *Pilea* species (bright indirect), page 47
- *Fittonia* species (humid), page 79
- *Tillandsia* species (humid), page 83
- *Haworthia* species (direct sun), page 32

Conclusion

Now that you have a better understanding of the environment you can offer your plants and the right plants for your space, I hope you'll use the skills and techniques we've covered here to care for your own plants. Remember that plant care is a journey, and while you will almost certainly make mistakes, you'll learn from them, gain greater confidence in your plant care skills, and see your houseplants grow happier and healthier.

Whether you're drawn to plants for what they bring to your space or what they bring to your state of mind, you can enjoy the connection they provide both to the natural world and to the people around you. As you care for them, let your love for your plants take root and nourish you in return.

RESOURCES

With so many places to find plants, planters, and other supplies for plant lovers, it can be overwhelming to sort through them all. In addition to the carefully chosen selection we carry at Folia (foliacollective.com), I've rounded up a few of my other favorite sources for you here.

BOOKS

While there are many wonderful plant books out there, these are a few of my tried-and-true choices for building general houseplant knowledge or diving deeper into the wonders of plants.

Success with House Plants by the Reader's Digest editors
This out-of-print book is fairly easy to find used and contains a wealth of easy-to-understand plant care information.

Root, Nurture, Grow by Caro Langton and Rose Ray
One of the best (and prettiest) books to get you started with propagation.

The House Plant Expert by Dr. D. G. Hessayon
A thorough reference book for all things houseplant.

What a Plant Knows by Daniel Chamovitz
An easy read that will open your eyes to many of the amazing things plants are up to.

The Hidden Life of Trees by Peter Wohlleben
A favorite of mine that will leave you looking at every tree in a whole new way.

Botany for Gardeners by Brian Capon
A wonderful botany primer written for non-scientists.

PLANT STUFF

Here are some brands making some of my favorite plant-related products.

GOOD DIRT
high quality potting mediums
good-dirt.com

MODERN SPROUT
stylish grow lights and more
modsprout.com

RT1 HOME
potting tarps, other plant supplies
rt1home.com

SOL TECH SOLUTIONS
minimal grow lights
soltechsolutions.com

WALLYGRO
wall pockets made for living walls and vertical gardening
wallygro.com

PLANTERS

Planters can often be found anywhere plants are sold, but finding stylish options made with plants in mind isn't as easy. In addition to vintage planters (and what we carry at Folia, of course), these are some of my favorite sources and makers of attractive homes for your plants that also take things like size and drainage into consideration and are available to order online.

CONVIVIAL
quality neutral designs that fit in anywhere
convivialproduction.com

GOPI SHAH CERAMICS
handmade ceramics with intricate carved designs
gopishah.com

JUNGALOW
fun assortment of planters in the bold bohemian style Jungalow is known for
jungalow.com

LIGHT + LADDER
stylish, minimal designs
lightandladder.com

LIGHTLY DESIGNS
lightweight spun metal designs in stunning colors
lightly.com.au

REVIVAL CERAMICS BY LBE DESIGNS
classic mid-century styles responsibly produced
lbedesign.com

TANDEM CERAMICS
classic shapes in a range of colors
tandemceramics.com

TOME CERAMICS
simple but interesting handmade designs
tomeceramics.com

PLANTS

I wholeheartedly encourage you to shop your local plant stores and nurseries, as well as find plant swaps or garden clubs to source new plants. If you're trying to track down a specific or unusual plant, though, these are a few shops I trust that sell plants online and offer an interesting selection with high quality standards.

ARIUM BOTANICALS
ariumbotanicals.com

GABRIELLA PLANTS
gabriellaplants.com

LOGEE'S
logees.com

NSE TROPICALS
nsetropicals.com

PISTILS NURSERY
pistilsnursery.com

ACKNOWLEDGMENTS

This book (and the speed at which it came to life) would not be possible without the wonderful community I'm blessed to be a part of.

First, to my always patient husband, **BILL**. Thank you for encouraging me, lending your keen editing skills as I was writing, helping out at the shop, and putting up with all the mess that days and days of photo shoots created in our home.

To my amazing team at Folia: **ARIANA**, **JACOB**, **LINDSAY**, **MAYA**, **MINDY**, and **STELLA**. Thank you for sharing your talents, kindness, and expertise with both me and the Folia customers every day; for holding down the fort at the shop so I could focus on this book; for assisting with these photoshoots; and for all the ways you make Folia a place of joy for me.

To the plant lovers whose homes appear in these pages. I'm so grateful you allowed me to come into your homes with my camera, or were kind enough to lend photos for those I couldn't visit in person. It's a privilege to have your beautiful spaces grace this book.

To my dear friends—thank you for encouraging me, supporting the shop, and extending me grace for disappearing while I was working on this project. Special thanks to **SUZ & TROY** for lighting a spark in me that would become Folia and for all the tangible ways you've shown me you believe in my ability to carry out the vision God has given me for this endeavor. To **JUSTINA**, for taking the time to coach me through the early stages of this book. To **ALEXA & ERIC** for lending your talents as filmmakers and for helping load and unload cars full of props over two days of shooting.

Thank you to **KIRSTEN** at A 1000 X Better and **CAROLINE** at Coco Carpets for the prop loans, and to **ANNE & CAROLINE** at Light Lab and **LAUREN** at LME Studios for the use of your beautiful studio spaces.

To the team at Girl Friday Productions without whom this book would not exist. To **KRISTIN MEHUS-ROE** for seeking me out; **EMILIE SANDOZ-VOYER** for guiding me through every step of this process; **CAREY JONES** for quick and insightful editorial skills that kept me on track; **MICAH SCHMIDT** for support with the studio shoots; and **PAUL BARRETT**, **KATY BROWN**, and **CHRIS NAVRATIL**—your ideas, expertise, and talents show in every page of this book.

To everyone at HMH, especially **SARAH KWAK**, **ALLISON CHI**, and **DEB BRODY**. It's an honor to see my name on a title from such an esteemed publisher.

Last, but certainly not least, to everyone in the plant-loving Folia community. Your support of this book and the shop means the world and literally helps us keep the lights on. Thank you for allowing me to grow Folia from that little pop-up shop to what we are today.

CONTRIBUTORS

HOMES

Name: Miki Carter
City: Pasadena, CA
Favorite plant: *Mimosa pudica* or any carnivorous plant
Website: linkedin.com/in/ mikicarter
Social media handle: @Plot.Twist.Interiors
Home appears on pages 52, 67, 98, 154

Names: Pop Annemarie Chan & Ash Chan; children Brandon, Lucas & Jacob; and kitties Oliver & Milo
City: Silverlake, CA
Favorite plant: *Dracaena marginata*
Website: wyldbnchplants.com
Social media handles: @Popannemarie79 @wyldbnchplants @thecontaineryard
Home appears on pages 25, 40

Name: Dabito
City: Los Angeles, CA
Favorite plant: bird's nest fern
Website: oldbrandnew.com
Social media handle: @dabito
Home and photo appear on page 75

Name: Deanna M. Florendo
City: San Francisco, CA
Favorite plant: *Hoya rotundiflora*
Social media handle: @habitpattern.sf
Home and photo appear on page 56

Name: Ron Goh
City: Auckland, New Zealand
Favorite plant: *Monstera*
Website: mrcigarloft.com
Social media handle: @mrcigar
Home and photos appear on pages ii, 24, 38

Names: Kristin & Gabriel Guy and pup Ginger
City: Pasadena, CA
Favorite plant: *Peperomia caperata*
Website: dinexdesign.com
Social media handle: @dinexdesign
Home appears on pages 3, 55, 86

Names: Libby Hockenberry & Cary Smith and cats Velvet & Isis
City: Atlanta, Georgia
Favorite plant: *Anthurium clavigerum* (Libby); *Copiapoa cinerea* (Cary)
Website: thevictorianatlanta .com
Social media handle: @thevictorianatlanta
Home and Cary's photos appear on pages 5, 28

Names: Christine & Jamie Kelso and pup Violet
City: Cardiff, CA
Favorite plant: currently *Rhaphidophora tetrasperma*
Website: workhardplanthard .com
Social media handle: @workhardplanthard
Home appears on pages 63, 96, 103, 119, 182

Names: Julien & Jonna Micoud, children Winston & Jacques, and Jake the Labrador
City: Escondido, CA
Favorite plant: All *Copiapoas* and columnar cacti in general (Julien); *Hoyas* and *Mammillaria cacti* (Jonna)
Social media handles: @earthwindandcactus @bonjourjonna @pardon_my_cactus
Home appears on pages 34–35, 90

Names: Mila Moraga-Holz & George Prior and children Victor, Max & Rex
City: Culver City, CA
Favorite plant: this week, rubber plant
Website: jestcafe.com
Social media handle: @mila_jestcafe
Home and Mila's photos appear on pages 39, 100–101

Names: Whitney Leigh Morris,
husband Adam, son West,
and rescue beagles StanLee
& Sophee
City: Venice, CA
Favorite Plant:
Variegated arrowhead vine
Website: TinyCanalCottage.com
Social media handle:
@whitneyleighmorris
Home and photo appear on
page 88

Names: Christy & Link Neal and
children Lily, Lincoln & Lando
City: Los Angeles, CA
Favorite Plant: lipstick vine
Social media handle:
@christyneal
Home appears on pages 29, 54,
60, 89, 148, 150

Names: Zoë & Neil Pearson,
daughter Nova, and two crazy
shelter cats
City: London, UK
Favorite plant: monkey mask
(*Monstera adansonii*)
Social media handle:
@convo_pieces
Home and Zoë's photos appear
on pages 7, 41

Name: Jeannie Phan
City: Toronto, Canada
Favorite plant: bird of paradise
Website: fieldnotesbystudioplants
.com
Social media handle:
@studioplants
Home and photo appear on
page 40

Name: Michelle Qazi
City: Bellevue, WA
Favorite plant: *Hoyas*
Website: 6thanddetroit.com
Social media handle:
@6thanddetroit
Home appears on pages
26, 68, 99
Photos of Michelle's home by
Christel Robleto

Names: Ariana Tanabe and
Gremlin the cat
City: Los Angeles, CA
Favorite plant: *Hatiora
salicornioides*
Social media handle:
@arianatanabe
Home appears on pages 71,
100, 181

Names: Sofie Vertongen &
Yannick De Neef and cats
Charlie & Josie
City: Antwerp, Belgium
Favorite plant: *Epiphyllum
anguliger* (Yannick); *Polyscias
balfouriana* (Sofie)
Website: theplantcorner.com
Social media handle:
@theplantcorner
Home and Sofie's photos appear
on pages 50–51, 91, 97

Names: Sara Weiss & Garrett
West and dogs Barry & Milah
City: Los Angeles, CA
Favorite plant: *Hoya linearis,
Monstera adansonii, Ficus elastica*
Social media handles:
@best.fronds.forever
@weezinbox
Home appears on pages 13, 27,
28, 39, 62, 144, 147, 180

SOIL RECIPES

Name: Hilton Carter
City: Baltimore, MD
Favorite plant: fiddle leaf fig
Website: thingsbyhc.com
Social media handle:
@hiltoncarter
Photo of Hilton Carter by
Aaron Keeny

Name: Enid Offolter
City: Plantation, FL
Favorite plant: *Anthurium
warocqueanum*
Website: nsetropicals.com
Social media handle:
@nsetropicals
Photo of Enid Offolter by Larry
Dieterich

Name: Jen Tao
City: Camarillo, CA
Favorite plant: *Echeveria* 'Lola'
and *Scindapsus pictus*
Social media handle:
@jenssuccs
Photo of Jen Tao by Becca
Stevens

Photo of baby Danae on page ix
by James DeBree

INDEX

ABOUT THE AUTHOR

DANAE HORST is the founder of Los Angeles–based plant boutique and plant styling studio, Folia Collective. Her life-long love of plants and interior styling, paired with her desire to empower people to keep their plants happy and healthy, led her to start Folia in 2016. With her background as a plant enthusiast, interior photo stylist, photographer, and writer, Danae has guided tens of thousands of people to the right plants for their space and lifestyle.

Filled with plants, stylish "plant stuff," and other goods for plant lovers, Folia has been named one of the "30 Cutest Plant Shops Around the World," and the Folia Instagram feed was called a "Plant-Filled Instagram That Will Turn Your Black Thumb Green" by Domino. Danae teaches regular workshops on a range of plant-related topics, and she has been featured in the *Wall Street Journal*, the *Los Angeles Times*, *BUST*, *House Beautiful*, and more.

Danae and her husband, Bill, live in a plant-filled bungalow in Pasadena, California.

danaerolynhorst.com   *foliacollective.com*   *@danaerolynhorst*   *@foliacollective*

ABOUT THE ILLUSTRATOR

RENEE GRIFFITH is a West Palm Beach-based designer and owner of HeartSwell, a thoughtfully illustrated stationery line and graphic design studio. She and Danae collaborated to bring Folia's branding to life. You can find Renee's plant illustrations throughout this book and in Folia's Los Angeles shop, as well as on several collaborative products the two worked on together. See more of Renee's work at heartswellco.com and @heartswellco on Instagram.